Electronically Scanned Arrays

Electronically Scanned Arrays

Robert J. Mailloux

ISBN: 978-3-031-00406-3   paperback
ISBN: 978-3-031-00406-3   paperback

ISBN: 978-3-031-01534-2   ebook
ISBN: 978-3-031-01534-2   ebook

DOI: 10.1007/978-3-031-01534-2

A Publication in the Springer series

SYNTHESIS LECTURES ON ANTENNAS #6

Lecture #6

Series Editor: Constantine Balanis, Arizona State University

Library of Congress Cataloging-in-Publication Data

Series ISSN: 1932-6076   print
Series ISSN: 1932-6084   electronic

First Edition
10 9 8 7 6 5 4 3 2 1

Electronically Scanned Arrays

Robert J. Mailloux
University of Massachusetts, Amherst, MA

SYNTHESIS LECTURES ON ANTENNAS #6

ABSTRACT

Scanning arrays present the radar or communications engineer with the ultimate in antenna flexibility. They also present a multitude of new opportunities and new challenges that need to be addressed. In order to describe the needs for scanned array development, this book begins with a brief discussion of the history that led to present array antennas. This text is a compact but comprehensive treatment of the scanned array, from the underlying basis for array pattern behavior to the engineering choices leading to successful design.

The book describes the scanned array in terms of radiation from apertures and wire antennas and introduces the effects resulting directly from scanning, including beam broadening, impedance mismatch and gain reduction and pattern squint and those effects of array periodicity including grating and quantization lobes and array blindness.

The text also presents the engineering tools for improving pattern control and array efficiency including lattice selection, subarrray technology and pattern synthesis. Equations and figurers quantify the phenomena being described and provide the reader with the tools to trade-off various performance features. The discussions proceed beyond the introductory material and to the state of the art in modern array design.

KEYWORDS

antenna array, phased array, scanning antenna, antenna design.

*To the women in my life; my sweet Marlene and
my lovely daughters Patrice, Julie and Denise,
and to my parents Joe and Nora who
have always been my inspiration.*

Contents

Preface

This lecture is an introduction to the most important topics that dictate the behavior of scanning array antennas. It is intended to address the needs of engineers familiar with electromagnetics, microwave technology and antennas or antenna systems.

The title of the book, Electronically Scanned Arrays employs the term "electronic scanning" rather than the more commonly used "phased" to emphasize that phase shifters and phase control are only one tool for array control. Of equal importance these days is the need for time delay devices and for active control of the amplitude distribution across the array. Much of this control can be done by analog phase shifters and switches, but future arrays will involve more use of digital and optical control where appropriate.

The study of array antennas includes the electromagnetics of radiation and the interaction of the various radiating array elements. It requires the understanding of a number of phenomena peculiar to arrays and not the individual elements. Among these are the grating lobes that result directly from periodicity and the many phenomena that result from scanning, like the broadening of the array radiated beam, the reduction of gain, the varying interaction between elements and resulting impedance changes. These and other phenomena characterize the behavior of scanning arrays.

The lecture begins by introducing the basic parameters of arrays and presents methods for calculating in its simplest form, giving some mathematical detail for basic dipole elements in Chapter 2 and arriving at the concept of an array element pattern. Chapter 3 introduces only a few basic techniques for pattern synthesis. This is a rich topic, but summarized here by several methods that are arguably the most important ones. The final chapter treats one of the most fundamental issues of array design, how to group elements together to save on phase shifters, time delay units or digital beamforming ports. These groupings are called subarrays, and seeking optimized methods of forming subarrays is of continuing interest in this environment of increasing bandwidth and increasing array size.

I am pleased to acknowledge the support of and many illuminating discussions with my colleagues at the Air Force Research Laboratory Electromagnetic Technology Division over the years, and throughout numerous institutional name changes. I thank my Air Force colleagues Allan Schell, Jay Schindler, Peter Franchi, Hans Steyskal, John McIlvenna, Jeff Herd, Boris

Tomasic, Livio Poles and David Curtis, and Arje Nachman of the Air Force Office of Scientific Research for the support of some of the fundamental aspects of antenna research.

Robert J. Mailloux
University of Massachusetts, Amherst, MA

CHAPTER 1

Basic Principles and Applications of Array Antennas

1.1 INTRODUCTION

The most important property of an antenna or antenna array is its ability to direct radiated power at some desired direction while suppressing it in all other directions. One measure of this advantage is the antenna realized gain $G(\theta, \phi)$. For a given input power at the antenna principles, the antenna realized gain is the ratio of the power density transmitted to some distant point relative to that of a lossless isotropic (omnidirectional) antenna that radiates equally in all directions. Figure 1.1 depicts an antenna pattern with directional gain, defines the coordinate system to be used later, and includes a sketch of an array of microstrip antenna elements.

The power density S_I (W/m^2) observed at some large distance R from a lossless isotropic (omnidirectional) transmitter antenna with input power P_0 at the antenna terminals is

$$S_I(\theta, \phi) = \frac{P_0}{4\pi R^2} \qquad (1.1)$$

and is just the radiated power distributed equally on the surface of a sphere.

When a directional antenna like that of Fig. 1.1 is used that power density becomes:

$$S(\theta, \phi) = \frac{P_0 G(\theta, \phi)}{4\pi R^2}. \qquad (1.2)$$

Here the quantity $G(\theta, \phi)$ is the antenna gain, and it is defined by the ratio of S to S_I. For a system that radiates other than a single linear polarization we use this expression to define the power density for each polarization. In the antenna's far field, a distance $R > 2D^2/\lambda$ for D the largest dimension of the antenna or array, the radiation propagates radially as a spherical wave and the power density is often expressed as a vector $\hat{\mathbf{r}} S(\theta, \phi)$(to indicate its radial direction), and called the Poynting vector $\mathbf{S}(\boldsymbol{\theta}, \boldsymbol{\phi})$.

Although antenna realized gain $G(\theta, \phi)$ is the primary measurable, and the term that is used in radar and communication range calculations, it is determined by first calculating a parameter called directivity that can be evaluated directly from the electromagnetic properties of the array. The directivity of an antenna is entirely related to the antenna pattern, not including

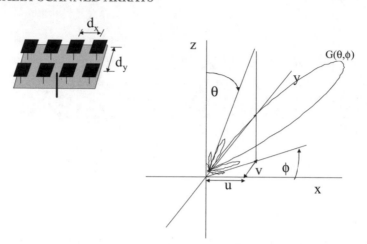

FIGURE 1.1: Array coordinates and power pattern

network, transmission line or even antenna losses. The directivity D_0 is defined as the ratio of the power density at the beam peak, divided by the total power radiated.

$$D_0(\theta, \phi) = 4\pi R^2 \frac{S(\theta, \phi)}{P_{\text{rad}}} = 4\pi \frac{S(\theta, \phi)}{\iint\limits_{\Omega} S(\theta, \phi)\text{d}\Omega} \qquad (1.3)$$

where Ω is the solid angle in steradians, and the integral is over 4π steradians.

Realized gain is related to directivity by the following relationship that accounts for loss and reflection within the antenna:

$$G(\theta, \phi) = (1 - |\Gamma|^2)\varepsilon_{\text{L}} D(\theta, \phi) \qquad (1.4)$$

where the reflection coefficient Γ is measured at the input terminals of the antenna, and ε_{L} is the dissipative loss in the array and associated transmission media. The term 'realized gain' differs from the IEEE standard definition of 'gain' that does not include reflection loss. Realized gain is the more relevant for array applications.

The directivity D_0 is a function of the antenna pattern only, so if the identical pattern is developed by an array or a single antenna, then the directivity is the same.

Although gain and directivity are used above as functions of the angular variables, these terms are also used to define the gain and directivity at the peak of the antenna beam, i.e., the maximum value of the terms. So we may speak of the gain of an antenna as being, say 30 dB, or speak of the gain of the same antenna in some certain direction as being 10 dB.

The advantage of providing a scanning beam instead of a stationary antenna pattern does not come without a price. Using a phased array instead of a single element or a larger antenna that does not scan is a matter of choice, but the overriding issue of cost vs. function will likely always dominate the selection process. Reflector and lens antenna systems can provide single or clustered multiple beam radiation patterns with rapid electromechanical scan. An attractive alternative, and sometimes lighter weight or less bulky, is the flat-plate array, basically a power divider feeding a broadside array, or a group of slotted waveguides or other single beam non-scanned array. These kinds of antennas are almost always less expensive than scanning arrays and almost always the choice unless scanning is actually required by the system specifications.

Phased arrays are a reasonable choice when they provide extra flexibility as required for modern radar and communication systems. These features, in order of the degree of sophistication, are: rapid electronic scanning in one or several planes, low sidelobes or flexible sidelobe control, multiple clustered scanning beams, adaptive pattern control against jammers or clutter, and fully independent low sidelobe multiple beams with each beam adaptively controlled.

These are capabilities that bring system designers to completely re-think the radar or communication system architecture. Arrays were often second choice in systems defined by older specifications, but many radar and communication systems now require the flexibility and added control afforded by arrays. There are a number of relevant texts [1–5] that describe array theory and technology.

1.2   A HISTORICAL PERSPECTIVE

Since the time of Marconi's paper entitled "Directive Antenna" [6] researchers have been investigating means for improving long-range communication with enhanced directivity. The history of radiating systems reveals a continuing search for increased antenna gain and narrower beamwidths that has gradually evolved into a search with more sophisticated goals. In the 1920s, demands for increased gain were satisfied primarily by fixed beam antennas. An example is the work of Beverage [7] who invented what he called a "wave antenna", now commonly called the "Beverage" antenna, to increase the gain above that of dipoles and other simple elements. The Beverage antenna is a long horizontal wire of about one wavelength mounted over earth, and loaded with appropriate resistances. The studies revealed choices of length, height, and terminating resistance to optimize the received signal. The popular Yagi [8] antenna is another antenna that uses multiple parasitic dipoles to establish a bound "surface wave" with enhanced gain.

Although radar was later to become the more significant impetus to array technology development, it was short wave radio that first utilized the flexibility of scanned arrays. As early as 1925, Friis [9] published a study of the reception of waves at 482 and 600 m wavelengths of several loop antennas whose signals were combined electronically, but not scanned. These loops were used to demonstrate the enhanced directivity of the antenna pair, and these were rigidly joined and rotated together to change the direction of the combined beam. Friis described a similar experiment at long wave ($\lambda = 5000-6000$ m) with square loops 8 ft on a side, and located 400 m apart. This larger structure was not rotated.

By the mid-1930s, numerous researchers had successfully combined radiating apertures or dipoles to form higher gain/higher resolution arrays. Increased directivity was only part of the solution however, since the received signal did not arrive from one elevation angle, but from waves which arrive at different vertical angles that change (although slowly) with time. Increased vertical directivity alone can eventually result in a pattern so narrow that the signal arrives outside of the angular range of the antenna much of the time. A more optimum solution was to form several beams and optimize the combination.

To address this issue, in 1937 Friis and Feldman [10] reported the results of a multiple unit steerable antenna (MUSA) for short wave reception. This antenna system supported a very sophisticated test of angle and time diversity combining. The test array consisted of six rhombic antennas arrayed along a three-quarter mile path in the direction of propagation. To accommodate the angle diversity (Fig. 1.2) the received signals were amplified and down-converted to an intermediate frequency where they were split to form three independent beams, each controlled in elevation by three sets of mechanically rotated phase shifters geared together to form the desired progressive phase Δ, 2Δ, 3Δ, etc. Of the three beams thus formed, one was used to search for the principal returns, and the other two were set at the angles with strongest signals. These outputs were then time delayed, passed through a network to equalize loss over the frequency band, and added to complete the combination of angle and time diversity. This first phased array achieved nearly 8 dB increase in average signal to noise power, relative to that of a single rhombic and proved the value of being able to select the optimum

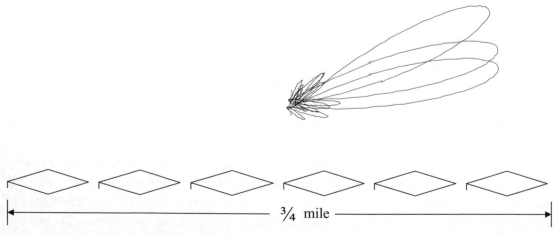

FIGURE 1.2: Three beam rhombic array of Friis and Feldman.

group of angles at any time then combine multiple time-delayed output signals. The three beam phasing and combining were done automatically using the fourth beam as monitor. This early demonstration of the value of electronic (or more accurately electromechanical) scanning certainly verified the electromagnetic foundations and at least one very practical application of the technology, but did not lead to fully electronic scan without many more years of technology development.

These communication-based applications pointed out the need for a technology that did not exist in the 1940s and even later, but they served to outline what the scanning array had to do. Phased arrays were essential to define and move narrow beams, to form independent multiple beams, to form low sidelobe radiating beams etc. The radar application brought in additional requirements for extremely rapidly scanning beams, wide band performance, the ability to adapt the beam patterns to avoid jammers and clutter. The first radar application was the British Chain Home radar [11]. That radar, built between 1935 and 1942, included a set of transmit arrays consisting of eight horizontal dipoles arrayed in the vertical plane and radiating broadside, and a set of four horizontal dipoles used to fill the elevation coverage gap where the eight element array had a null.

In April 1947, Friis and Lewis [12] published a detailed account of radar antenna technology including the basic electromagnetics of antenna radiation and the effects of amplitude taper and linear, square law, and cubic aperture phase variation. They included a short catalog of military land based and airborne radar antennas, most of which were basic apertures, like parabolas or metal plates lenses but including some details of an electromechanically controlled phased array, the Polyrod Fire Control antenna developed by Bell Laboratories and based on the MUSA experience [12–14]. This array, shown in Fig. 1.3, consisted of 14 S-band elements, with each element an in-phase vertical array of three polyrods. The inter-element spacing was about two wavelengths since the array was only required to scan $\pm15°$ in azimuth. The high gain and narrowed pattern of the polyrod enabled the two wavelength spacing and reduced element count while suppressing some of the grating lobes. Rotary phase shifters invented by Fox [13] provided a reliable mechanical phase shift, but not truly electronic scan. The phasing was accomplished by an ingenious arrangement of cascaded rotary phase shifters as shown in Fig. 1.4. The array was center-fed and used extra fixed line lengths to equalize phase across the array with all phase shifters set to zero. The phase shifters were ganged together and scanned the beam continuously with inter-element progression of 3600°/s. This system was successfully used in Word War II as the Mark 7 and Mark 8 fire control radars. Figure 1.5 shows the Mark 8 radar (along with two others) that was put into production in 1942 by Western Electric Company (a Bell Laboratory subsidiary). Additional details about this array are given by Kummer [14] and by Fowler [15].

FIGURE 1.3: Polyrod MUSA (Multiple unit scanning array).

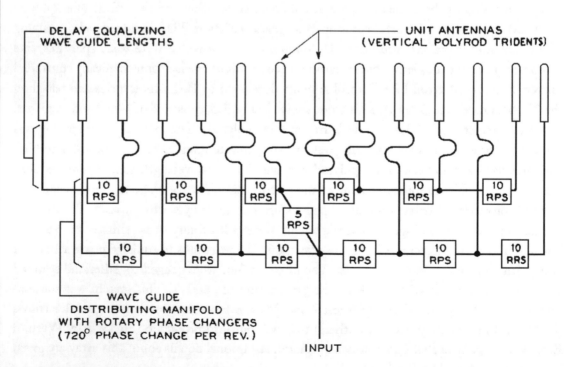

FIGURE 1.4: Phase control for the MUSA array.

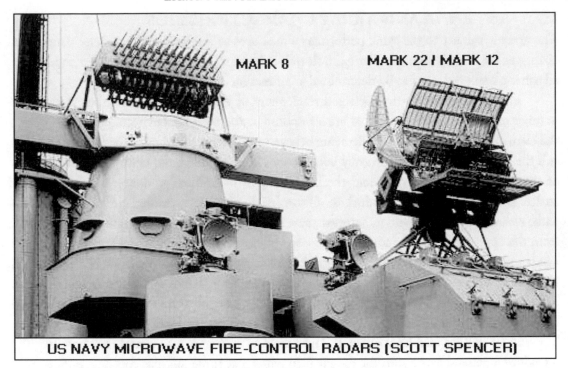

FIGURE 1.5: Mark 8 shipboard fire-control radar.

Another early scanning array was the German Wullenweber array. The basic Wullenweber is a set of vertical dipoles arranged in a circle with receiving ports connected to a sector of the array, and that sector commutated around the circle by means of a capacitive coupling. The received signals were fed to a radiogoniometer [16] that combined the signals and produced a response proportional to angular deflection. The device was used for azimuth direction finding.

The success of the polyrod and Wullenweber antennas still did not lead to any obvious transition to the technology of today. The lack of an electronic phase shifter was to limit scanning systems to electromechanical scan or frequency scan until the later invention of the microwave ferrite phase shifter. The impossibility of cascading large numbers of Fox rotary phase shifters led to ingenious electromechanically scanned mechanisms, and some of these are documented in the chapter by Kummer [14] and the article by Fowler [15].

These historical developments uncovered many of the basic principles of array theory and technology. Phase and time delay scanning, multiple beams and even a degree of adaptive control were all well understood by the early 1940s. It was another 20 years before the technology of electronic control brought us the phase shifters, time delay devices and digital control to begin to explore the full potential of scanning array antennas.

1.3   ANTENNA AND ARRAY CHARACTERISTICS

The array is subject to the same performance measures as any antenna, but these parameters change as the array is scanned. The gain, sidelobes, input standing wave ratio, polarization, and all other parameters need to be determined as a function of scan.

The array elements themselves dictate many of the array properties, but there are a number of array characteristics that are a response to the periodic environment that supports the elements. Although isolated elements behave very differently when embedded in an array, still the elements determine the array polarization, fundamental bandwidth and gross features of the array radiation pattern, and the inter-element or 'mutual' coupling within the array lattice. Basic array elements can include all sorts of dipoles, slots, printed circuit or microstrip patch radiators and wide band radiators that are flared notch elements and mounted to protrude from the aperture. The only condition imposed by the array is that spacing between elements needs to be small enough to avoid the grating lobes and "blindness" that will be discussed in Chapter 2. This requires that one must assume spacing of less than one half wavelength at the highest frequency for wide angle scan, and so the elements must be relatively small. Detailed descriptions of radiators suited to be array elements are in the literature [17–19].

More will be said later about the near field of elements and arrays, but here we first introduce a general expression for the far field pattern at some angular coordinate (θ, ϕ) of a single element in the array. The array elements are at coordinates (x', y', z'), and the excitation of the nth element is a complex number a_n:

$$\mathbf{E}(u, v) = c \frac{e^{-jkR_0}}{R_0} \mathbf{f}(u, v) e^{j\mathbf{k}\cdot\mathbf{r}'} a_n, \qquad (1.4)$$

where

$$u = \sin\theta\cos\phi, \quad v = \sin\theta\sin\phi, \quad \mathbf{k} = k(\hat{x}u + \hat{y}v + \hat{z}\cos\theta)$$

$$k = \frac{2\pi}{\lambda}, \quad \mathbf{r}' = \hat{x}x' + \hat{y}y' + \hat{z}z', \quad \text{and} \quad R_0 = \sqrt{x^2 + y^2 + z^2}$$

In this expression, the constant c pertains to the kind of element used, and is evaluated directly or, more often, expressed by a relationship to element gain. The array element pattern $\mathbf{f}(u, v)$ is a vector quantity, is often normalized to unity at its peak except when being used to evaluate gain, and includes all polarization effects. The parameters u and v are the direction cosines of a spherical coordinate system in (r, θ, ϕ).

The array element pattern for any element in the array is measured far from the array by driving that element and match-loading all other elements exactly as in measuring the network scattering matrix. Since all the elements in an array are coupled, the element pattern $\mathbf{f}(u, v)$ includes the contributions of all elements in the array, not just the element at (x', y', z'). Any particular array element pattern will be different from that same element in a different location in the array and different from the same element not in the array. In practice, it is usually the elements near an array edge that behave most differently.

It is often convenient to omit the constant $c(e^{-jkR_0}/R_0)$ since this is just a constant multiplier.

The pattern for the array of elements is thus:

$$\mathbf{E}(u, v) = \sum_n a_n \mathbf{f}_n(u, v) e^{j\mathbf{k} \cdot \mathbf{r_n}'} \qquad (1.5)$$

where the equation accounts for the fact that all element patterns $\mathbf{f}_n(u, v)$ may be different.

In order to simplify these general expressions, and to describe the most important array parameters in the simplest terms, we will restrict the analysis to arrays of elements lying in a plane and arranged periodically in one or two dimensions as in Fig. 1.6.

We first consider a linear array of N equally space elements arranged in a line and spaced d_x apart as shown in Fig. 1.6. We will assume that all element patterns are the same, and so will not be included in what follows. This is a practical assumption, and will be discussed further in Chapter 2.

In this case $\mathbf{r}' = \hat{\mathbf{x}}x' = \hat{\mathbf{x}}nd_x$ and the resulting pattern is:

$$\mathbf{F}(u, v) = \mathbf{f}(u, v) \sum_{n=-(N-1)/2}^{(N-1)/2} a_n e^{jknd_x u} = \mathbf{f}(u, v) F_A(u) \qquad (1.6)$$

In this form, the expression shows the array pattern as the product of a vector element pattern and a scalar algebraic term $F_A(u)$ which we call the array factor.

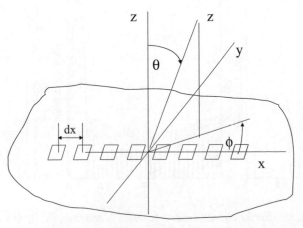

FIGURE 1.6: Linear array and spherical coordinates.

1.3.1   Array Scanning: Linear Array

The excitation coefficient a_n can be chosen to make the pattern peak always point in the direction $(\theta_0, 0)$ for all frequencies, so that $u_0 = \sin\theta_0$. By selecting the excitation a_n to be a time-delayed signal:

$$a_n = |a_n|\, e^{-jknd_x u_0} \tag{1.7}$$

the array factor F_A is written for a time delay steered array

$$F_A = \sum_{-(N-1)/2}^{(N-1)/2} |a_n| e^{+jknd_x(u-u_0)} \tag{1.8}$$

that has its maximum value at $u = u_0$ at all frequencies. As the frequency is changed this pattern (see Fig. 1.7) broadens at lower frequencies and narrows at higher frequencies, but the peak is stationary because the excitation a_n is time delayed. The element spacing is 0.5 wavelength at center frequency, Fig. 1.7(a) is at one half of the center frequency, and u_0 is 0.866 (60 degree scan). Figure 1.7 also shows that if the frequency is increased too much, there may be an additional beam or beams that radiate due to the electrically large spacing between elements. These are called 'grating lobes' and discussed in a later section. Notice that aside from the grating lobes, the sidelobes are unchanged with frequency.

Expression 1.7 can be written in a form to highlight its time dependence.

$$a_n = |a_n|\, e^{-j\omega\tau_0} \tag{1.9}$$

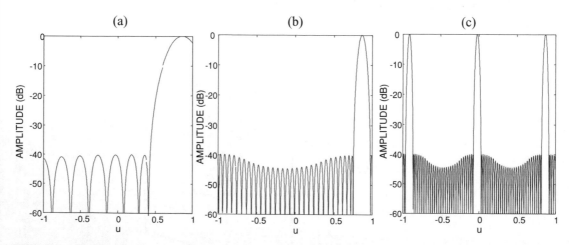

FIGURE 1.7: Array factors for 32 element array. (a) 0.25 λ spacing, (b) 0.50 λ spacing, and (c) 0.75 λ spacing.

where $\tau_0 = (n d_x u_0)/c$ is the time for light to travel the distance $n d_x u_0$. The exponent of Eq. (1.9) is the phase shift of a length of transmission line.

Although time delay offers "perfect" control, time delay devices have traditionally been switched coaxial transmission lines, and from (1.9) it is clear that they must be nearly as long as the array $(N d_x)$, and this may make them prohibitively expensive, lossy and too bulky for many applications. One can minimize the bulk of these lines by organizing the time delays in a 'tree' configuration, so that longer delays feed groups of elements fed by shorter delays that feed smaller groups of elements, etc., but this organization needs power amplifiers within the tree to make up for the significant loss. The alternative has been to use phase steering or phase steering within time-delayed subarrays, a subject to be discussed later.

1.3.2   Array Pattern Shape and Beamwidth

In an important special case, all excitation amplitudes are the same and the summation of Eq. (1.8) can be done explicitly. That pattern is written below for an array of N time-delayed elements, normalized to the maximum value.

$$F_{\mathrm{A}} = \frac{\sin\left(\dfrac{N\pi d_x}{\lambda}(u - u_0)\right)}{N\sin\left(\dfrac{\pi d_x}{\lambda}(u - u_0)\right)} \qquad (1.10)$$

This pattern is shown in Fig. 1.8(a). For a 32 element array of half-wave spaced elements scanned to $u_0 = .866$, corresponding to $60°$ scan has nulls at $u - u_0 = n\left(\frac{\lambda}{L}\right)$ for $L = N d_x$,

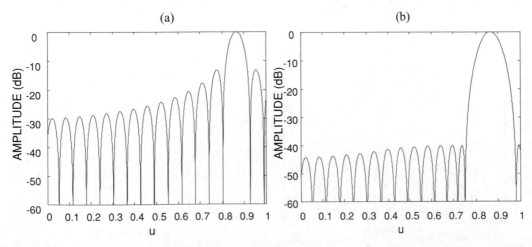

FIGURE 1.8: Array patterns for 32 element array with half wave spacing (a) p-sinc pattern of uniform array scanned to $60°$ and (b) Taylor 40 dB sidelobe pattern.

a dimension one element longer than the array. This pattern the p-sinc (or periodic sinc) function, and is very similar to the sinc pattern, which is the far field pattern of a uniformly illuminated continuous radiating aperture. The p-sinc pattern is different in that it has grating lobes for large arguments. Equations (1.6) and (1.8) show that as the array is scanned, the array factor does not change if plotted in u_0, v_0 space. Changing u_0 and v_0 simply translates the pattern but does not change its beamwidth or sidelobes. The actual radiated pattern in (θ, ϕ) space does change due to the mapping of (u, v) space into (θ, ϕ) space. Figure 1.9 shows a comparison of the same pattern of a linear array plotted in $u = \sin\theta$ space vs. θ space. The two figures at left show the broadside beams in the two scales, and the figures at right show that in angle space the beams broaden as the array is scanned. Later we show that for a planar array the gain is commensurately reduced with scan.

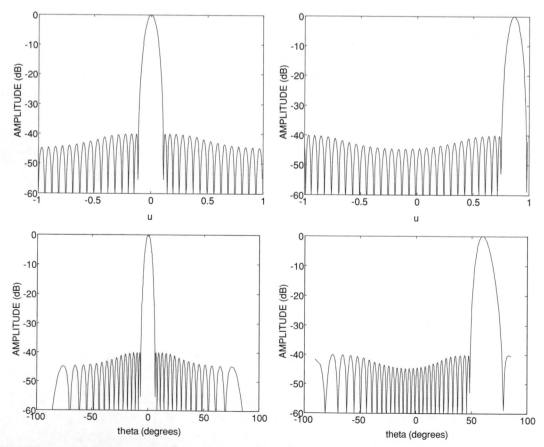

FIGURE 1.9: Array beam broadening with scan: Upper patterns are broadside and 60° scan in u-space. Lower patterns are same in angle space.

Except for certain "superdirective" patterns that can be synthesized for $d/\lambda < 0.5$ this pattern has the narrowest beamwidth [20] (and resulting highest gain) that can be produced with any set of excitation a_n. It also has relatively high sidelobes at −13.2 dB and others decaying asymptotically like $1/(n + .5)$ for large n. The "superdirective excitations" will not be discussed further as they are practical for only a few very specialized circumstances, and only for arrays of a few elements [21].

Other chosen excitations can have lower (or higher) sidelobes, and the judicious selection of these distributions is the subject of a later chapter. In general, lower sidelobe levels are obtained by 'tapering' the excitation amplitude distribution so that the largest weights are in the center of the array, with the weights gradually being reduced toward the array edges. The figure at right in Fig. 1.8(b) is a Taylor pattern [22] with −40 dB sidelobes.

An expression for the beam broadening of a periodic linear array is given in many references, but a more convenient approximate expression can be obtained for a large array scanned not too near to the "endfire" direction $\theta = 90°$. The beamwidth for a linear array or a two dimensional array varies approximately as:

$$\theta_3 = \frac{\theta_3^\circ}{\cos \theta} \tag{1.11}$$

where θ_3 is the scanned beamwidth along the direction of scan, and θ_3° is the beamwidth at broadside.

For a large uniformly illuminated array the half-power beamwidth of this array factor is approximately:

$$\theta_3 = 0.886 \frac{\lambda}{L} \frac{1}{\cos \theta} \tag{1.12}$$

In the case of a tapered array the beamwidth is broader than that of Eq. 1.12, and it is often convenient to reference the beamwidth to this basic p-sinc pattern by defining a beam broadening factor B_b, such that for a tapered array we write

$$\theta_3 = 0.886 B_b \frac{\lambda}{L} \frac{1}{\cos \theta} \tag{1.13}$$

1.3.3 Array Pattern 'Squint' and Bandwidth

Although the equations for time-delayed arrays were introduced earlier, most scanning arrays are "phased arrays" since the scan control is done using phase shifters before each element. In expression 7 this means setting

$$a_n = |a_n| e^{-j \frac{2\pi}{\lambda_0} n d_x u_0} \tag{1.14}$$

for some central wavelength λ_0. This makes expression 1.6 take the form below:

$$F_A(u) = \sum_{-(N-1)/2}^{(N-1)/2} |a_n| e^{jk_0 n d_x(\frac{f}{f_0} u - u_0)} \qquad k_0 = 2\pi/\lambda_0 \qquad (1.15)$$

This is identical to expression 1.7 at center frequency, but at other frequencies the radiated beam has its peak at $u = (f_0/f)u_0$. Figure 1.10 shows patterns of a linear array of 128 elements with phase steering as the frequency of operation is taken between f_0 and $1.013 f_0$ where the original beam (here shown solid) at f_0 and $\sin\theta = 0.707$ (45°) has moved to the dashed curve at $\sin\theta = 0.698$. This is the position at which the required signal will be 3 dB suppressed as compared to its value at the frequency f_0

As shown in the figure, the phase steering makes the beam peaks move toward broadside ($u = 0$) for $f > f_0$, and at frequencies below center frequencies to move away from broadside. The radiated signal thus wanders of the target as frequency is changed, and this imposes a very restrictive bandwidth limitation on phased steered arrays.

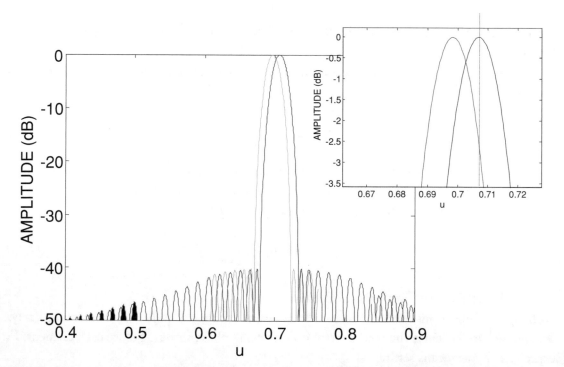

FIGURE 1.10: Array pattern of 128 element array scanned to 45° and drawn at center frequency (solid, $f/f_0 = 1.0$) and at band edge (dashed, $f/f_0 = 1.013$). Inset shows "squint" for case of 3 dB loss.

One can show that for 3 dB loss in signal reception due to this 'squint' the system bandwidth is:

$$\frac{\Delta f}{f_0} = \frac{\Delta u}{u_0}$$

(1.16)

where Δu is the array beamwidth measured at the −3 dB points.

Scanning a Planar Array

For a two dimensional array with elements at $x' = md_x$ and $y' = nd_y$, the basic equation (5) is unchanged, but using a double summation over the weighting a_{nm} with

$$\mathbf{r}' = \hat{\mathbf{x}}x' + \hat{\mathbf{y}}y'$$

(1.17)

The resulting array factors are:

$$F_A = \sum_{m=-\left\{\frac{M-1}{2}\right\}}^{m=\left\{\frac{M-1}{2}\right\}} \sum_{n=-\left\{\frac{N-1}{2}\right\}}^{n=\left\{\frac{N-1}{2}\right\}} |a_{mn}| \; e^{jk_0[md_x(\frac{f}{f_0}u-u_0)+nd_y(\frac{f}{f_0}v-v_0)]}$$

(1.18)

for phase steering and

$$F_A = \sum_{m=-\left\{\frac{M-1}{2}\right\}}^{m=\left\{\frac{M-1}{2}\right\}} \sum_{n=-\left\{\frac{N-1}{2}\right\}}^{n=\left\{\frac{N-1}{2}\right\}} |a_{nm}| \; e^{jk[md_x(u-u_0)+nd_y(v-v_0)]}$$

(1.19)

for time delay steering.

1.3.4   Grating Lobes of Linear Arrays

The patterns of Figs. 1.8–1.10 have a single beam since the element separation was chosen to be 0.5 wavelengths at the highest frequency. In fact, these arrays can have additional radiated beams, fully as large as the primary beam if the array grid dimensions are chosen too large. An example (Fig. 1.11) shows an array factor for a 32 element linear array with spacing 0.75λ. Here the beam is shown at broadside in Fig. 1.11(a) and scanned to 60° ($u_0 = .866$) in Fig. 1.11(b). The scanned beam is accompanied by another, called a grating lobe at about −40°. Note that this array factor pattern has been plotted in the angular space, and that highlights the fact that the beamwidth in angle space is not constant as the array is scanned.

In the case of a time-delayed linear array at frequency f, the grating lobe condition can be seen from the exponent of Eq. (1.7), which has the value zero for every u_p satisfying:

$$2\pi \frac{d_x}{\lambda}(u_p - u_0) = 2\pi p \quad \text{for } p = 0, \pm 1, \pm 2, ..$$

(1.20)

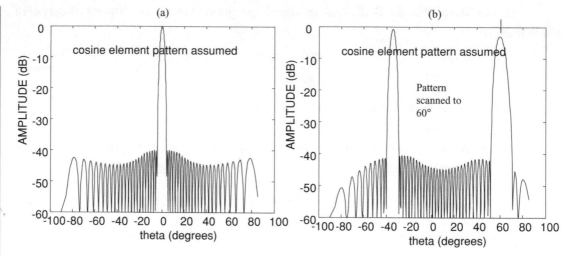

FIGURE 1.11: Array pattern for 32 element array with 0.75 wavelength spacing. Pattern at left shows broadside radiation with no grating lobes, while one at right shows a grating lobe near −40° larger than main beam at 60°

or when:

$$u_p = u_0 + \frac{p\lambda}{d_x}.$$

These values of u are the primary or main beam ($p = 0$) and a set of beams for all values of p and displaced from the main beam by the distance $p\lambda/d_x$ for any $p \neq 0$. Note that for a phase scanned array the main beam peak would be at $u = u_0(\lambda/\lambda_0)$ and so the grating lobes would be at

$$u_p = u_0\frac{\lambda}{\lambda_0} + \frac{p\lambda}{d_x}. \qquad (1.21)$$

Figure 1.12 shows an array factor for a linear array of widely spaced elements spaced 1.5 wavelengths apart with primary beam at $u_0 = 0.866$ (60°) and additional beams at $0.866 + (\lambda/1.5)p$. Two grating lobes radiate in the region $-1 \leq u \pm 1$, which (for $\varphi = 0$) is the region in which $|\sin\theta| \leq 1$ and often called 'real space'. This interpretation of 'real space' and the implication that there is 'imaginary space' for which $u > 1$ will be extended later to the planar array case, and it offers a way of looking at the solution as restricted to real angles.

Figure 1.13(a) gives a picture of the real-space/imaginary space boundary, and the grating lobe locations (circles) for a one dimensional array. The main beam is shown as a filled-in circle. One can readily show that all such grating lobes can be suppressed by selecting the spacing d_x

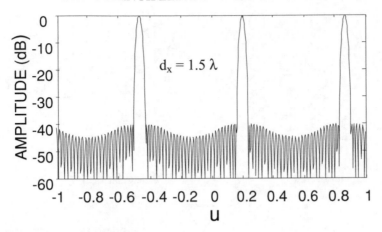

FIGURE 1.12: Pattern for 32 element array with 1.5 wavelength spacing showing multiple grating lobes

such that

$$\frac{d_x}{\lambda} \leq \frac{1}{1 + \sin \theta} \qquad (1.22)$$

for scan to some angle θ. This small spacing assures that if the radiating beam is chosen anywhere $|u_0| \leq 1$, then all other beams will be in imaginary space $|u_0| \geq 1$, and so not radiate.

Note that a small array might have a broad enough grating lobe beam that a grating lobe might have its peak in imaginary space but still have sizeable contribution in real space, so in the case of a relatively smaller array it is sometimes useful to choose the spacing d_x such that the grating lobe is one-half of the array half power beamwidth ($\sin \theta_3$) beyond the imaginary

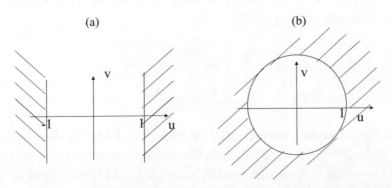

FIGURE 1.13: Real space and imaginary space: (a) Linear array and (b) planar array.

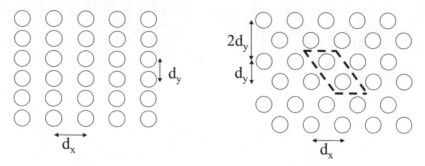

Rectangular Lattice   Triangular Lattice

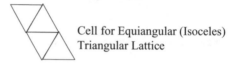

Cell for Equiangular (Isoceles)
Triangular Lattice

FIGURE 1.14: Planar array configurations: (a) Rectangular lattice and (b) Equilateral triangular lattice.

space boundary.

$$\frac{d_x}{\lambda} \leq \frac{1}{1 + \sin\theta + \sin\theta_3/2} \qquad (1.23)$$

Figure 1.13(b) shows the exclusion zone for a periodic planar array. The real space is bounded by a circle since now $\sin\theta = \sqrt{u^2 + v^2} \leq 1$. Figure 1.14 shows two types of grids for planar arrays, a rectangular grid and triangular grid (in this case an equiangular triangular) lattice.

1.3.5   Grating Lobes of Planar Arrays

Expressions for the grating lobe locations of an array with rectangular grid are similar to those for the linear array, and can be derived readily using the same logic. For a phase scanned array they are located at positions (u_p, v_q) with:

$$v_q = \frac{\lambda}{\lambda_0}v_0 + \frac{q\lambda}{d_y} \ . \qquad (1.24)$$

$$u_p = \frac{\lambda}{\lambda_0}u_0 + \frac{p\lambda}{d_x},$$

These grating lobes are shown as small circles in Fig. 1.15(a) and are again bounded by the real-space circle at $\theta \leq 90°$. The small filled in circle is the main beam at (u_0, v_0). For the rectangular grid array the nearest grating lobes occur in the principal planes, so the condition of Eq. (1.21) gives the bounds for both d_y and d_x.

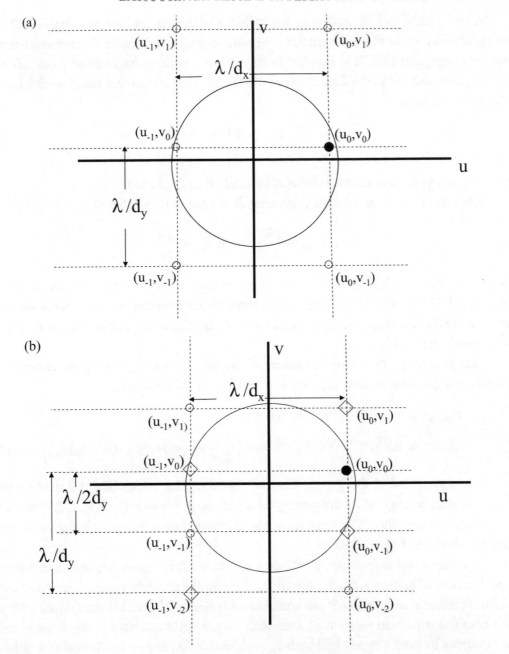

FIGURE 1.15: Grating lobe circles: (a) planar rectangular lattice and (b) planar equilateral triangular lattice.

A commonly used alternative geometry is the triangular grid array of Fig. 1.14(b). The grating lobe features of this configuration bear some explanation because they are more complex than the rectangular grid. It is simplest to describe as a rectangular array of paired elements, with the pairs outlined in the dashed box of this figure. This means that the time-delayed array factor can be written:

$$f(u, v) \sum_m \sum_n |a_{n\ m}| e^{j\frac{2\pi}{\lambda}\left[nd_x(u-u_0)+2md_y(v-v_0)\right]} \tag{1.25}$$

where f is the pattern of the two elements blocked off in the figure.

Since the array is time delayed, this array factor has grating lobes at

$$u = u_0 + \frac{p\lambda}{d_x}; \quad v = v_0 + \frac{q\lambda}{2d_y} \tag{1.26}$$

as shown by the small circles and diamonds on the figure. Notice that if the dimensions are not changed, this lattice has grating lobes spaced closer together in the v plane due to the periodicity of $2d_y$ that has been introduced by displacing the rows. These extra lobes lie on the added lines in Fig. 1.15(b).

The pattern f, the element pattern of the two element groups (also assumed time delayed), and given in the next expression, removes some of these lobes.

$$f(u, v) = 1 + e^{j\frac{2\pi}{\lambda}\left[(u-u_0)d_x/2+(v-v_0)d_y\right]}$$
$$= 2e^{j\frac{\pi}{\lambda}\left[(u-u_0)d_x/2+(v-v_0)d_y\right]} \cos\left\{\frac{\pi}{\lambda}\left[(u-u_0)d_x/2+(v-v_0)d_y\right]\right\}. \tag{1.27}$$

When evaluated at the grating lobes of the triangular lattice (Eq. 1.26), the zeros of the cosine pattern align with the grating lobe indices $p + q = \pm(1, 3, 5\ldots.)$, and this removes those grating lobes. The locations of the eliminated quantization lobes are indicated by diamonds shown in Fig. 1.15(b).

This allows one to decrease d_y but increase d_x to result in a geometry that takes advantage of the location of grating lobes for the triangular grid. In an early paper Sharp [23] showed that for a conical scan area of $45°$, an equilateral triangular (also called hexagonal) grid array, set to have the same scan limits as an array with a square grid can have 13.4% fewer elements. In an example he used a square grid with $d_x = d_y = 0.585\lambda$, and an equilateral triangular grid of the same d_x dimension, but $d_y = 0.338\lambda$. Other scan combinations, for example those with elliptical instead of circular scan areas have lesser advantage, or even none. So the triangular grid can be an advantage, depending upon the required scan coverage area. In general, one should consider both triangular and rectangular grid configurations if scan is to be optimized in one plane and not the other.

Figure 1.14(b) also shows why the equilateral triangular grid saves 13.4% of elements as compared to a square grid. The parallelogram outlined is the area occupied by two elements, and one can see that each element has the area of two equiangular triangles shown in the figure. The height of each is $d_y = \sqrt{3}\,(d_x/2)$, and since the area of each triangle is $\sqrt{3}(d_x^2/4)$, then the area $(d_x^2/0.866)$ occupied by the element cell is larger than the area (d_x^2) occupied by the rectangular grid with the same d_x spacing. The fully filled array thus uses 13.4% fewer elements than the equivalent rectangular grid array.

1.3.6 Array Directivity Equations

The directivity of a linear array of isotropic elements with half wave spacing (or any multiple of half wave spacing) is [24]:

$$D_0 = \frac{\left|\sum a_n\right|^2}{\sum |a_n|^2} \tag{1.28}$$

and has the value N for a uniformly illuminated array of N elements. The directivity is independent of scan for this isotropic case since the pattern in the plane orthogonal to scan is restricted by the real-space/imaginary-space boundary as the array is scanned toward endfire. The ratio of this directivity to the maximum directivity (N) is called the taper efficiency $\varepsilon_T = D_0/N$. Other expressions for the directivity of tapered linear arrays are given in the literature [21].

The directivity of a planar array can be shown to have a maximum of:

$$D_{\max} = \frac{4\pi A}{\lambda^2} \tag{1.29}$$

where the array area A is independent of the shape of that area.

One can also show [25] that for a very large array, since one can assume that all the array element patterns (except for a small number near the array edges) are the same, then the approximate maximum array gain can be written in terms of the gain of a measured element pattern $g_e(\theta, \phi)$.

$$G = \frac{|g_e|_{\max}\left|\sum_m \sum_n a_{mn}\right|^2}{\sum_m \sum_n |a_{mn}|^2} \tag{1.30}$$

where the measured element pattern includes all network losses. Equation (1.30) becomes an expression for the directivity if the losses are removed, and then the element pattern has a maximum value of $|g_e|_{\max} = \frac{4\pi}{\lambda^2} A_{\text{cell}}$, where A_{cell} is the total array area divided by the number of array elements. This expression reduces to $D_{\max}$ when the array is uniformly illuminated, and so the taper efficiency (aperture efficiency) is again given as:

$$\varepsilon_T = \frac{\frac{4\pi A_{cell}}{\lambda^2}}{\frac{4\pi A}{\lambda^2}} \frac{\left|\sum_m \sum_n c_{mn}\right|^2}{\sum_m \sum_n |c_{mn}|^2} = \frac{1}{N} \frac{\left|\sum_m \sum_n c_{mn}\right|^2}{\sum_m \sum_n |c_{mn}|^2} \qquad (1.31)$$

1.4   MULTIPLE BEAM NETWORKS

The ability to form several or even a multiplicity of beams with a single array is highly desirable. We have already discussed the advantage of multiple beams for the purpose of long-range communication, but they are also extremely useful for radar applications where they are used for beam shaping and as an aid to tracking targets. Figures 1.16–1.18 show three multiple beamforming systems that have been widely used.

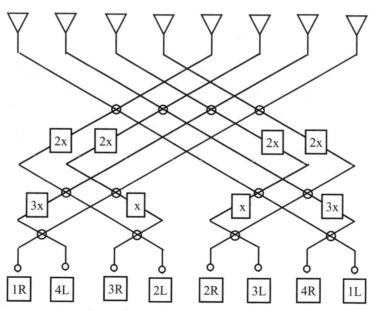

x = π/8 radians phase shift

⊠   Hybrid coupler convention: Straight through arms have no phase shift, while coupled arms have 90° phase shift

8 x 8 Butler Matrix

FIGURE 1.16: Butler matrix beamformer.

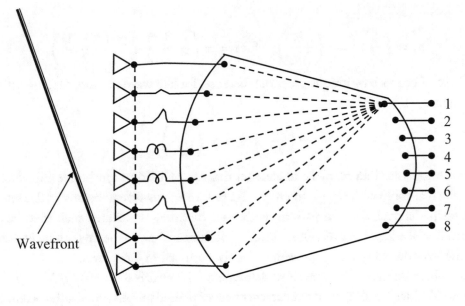

FIGURE 1.17: Rotman Lens beamformer.

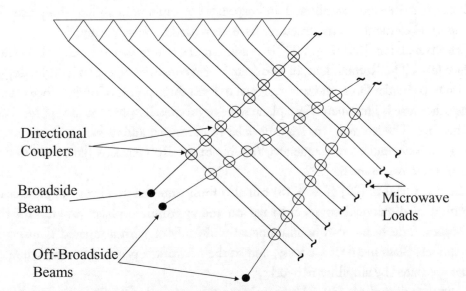

FIGURE 1.18: Time-delayed version of Blass Matrix beamformer.

The Butler matrix [25] of Fig. 1.16 is a classic network that radiates a number of beams (simultaneously) with each beam excited from a single input port. Each input port excites a set of uniformly illuminated array element ports with a progressive phase across the array. The array therefore radiates the periodic sinc (p-sinc) beams of Fig. 1.8(a), which have peaks at direction cosine coordinates

$$u_i = \left(\frac{\lambda}{L}\right) i = i \left(\frac{\lambda}{Nd_x}\right) \quad \text{for } i = \pm \left(\frac{1}{2}, \frac{3}{2}, \frac{5}{2}, \ldots, \frac{(N-1)}{2}\right). \qquad (1.32)$$

These N beams have unique properties because, for half wave spacing, they are orthogonal over the range $-1 \leq u \leq 1$.

$$-\left(\frac{\lambda}{2d_x}\right) \leq u \leq \left(\frac{\lambda}{2d_x}\right)$$

Taken together they form a complete set that can be used to synthesize any pattern that is physically realizable with the given array. Figure 1.16 shows several beams in the set.

A further convenience for pattern synthesis is that they are located such that the peak of each beam is at the zeros of every other beam. This means that the beams can be conveniently used in the Woodward synthesis procedure that is discussed in Chapter 3.

The phase difference between any element and the next is $\delta_i = i(2\pi)/N$ for an N-port Butler matrix. Since the Butler matrix presents a set of phased output signals, the beams formed are not time delayed, and so the output beams squint with frequency.

Butler matrices are complicated and expensive to build, and so have only been used to control up to 64 elements. Lens structures have been used for larger arrays.

The Rotman lens [26] (Fig. 1.17) is a variation of a lens invented by Gent [27] and called a 'Bootlace lens'. The Rotman lens, in its original form consisted of a set of conducting parallel plates with two circular faces, and was designed to form three points of perfect focus at one face when the other face is illuminated by a plane wave or progressive phase at the appropriate angle at the other face. The Rotman lens is a time delay beamformer, and its beams are approximately fixed in space independent of frequency. Variants of the Rotman lens have been designed in stripline and microstrip media [28].

The Blass Matrix [29] (Fig. 1.18) can also form time-delayed beams, and uses a series configuration of directional couplers to tap off the appropriate phase progression for each radiated beam. The beams can be illuminated uniformly or with a tapered illumination for lower sidelobes. Blass matrices are lossy, and so the network is primarily useful when the loss can be compensated by amplifier networks.

A number of authors [30, 31] have shown that one cannot build a lossless network to radiate more than one independent beam form an aperture except when certain very specific conditions are met. These conditions are related to the shape of the radiation patterns and the spacing between the beams. They amount to a statement that the beams cannot be formed by a lossless network unless they are orthogonal.

Stein [32] gives a very useful limit to the maximum efficiency of a network radiating two beams. His expressions are summarized in the text [2], and so will not be reproduced here.

REFERENCES

[1] R. C. Hansen, *Microwave Scanning Antennas*. New York: Academic Press, 1964.

[2] R. J. Mailloux, *Phased Array Antenna Handbook*, 2nd edition. Boston, MA: Artech House Inc., 2005

[3] R. C. Hansen, *Phased Array Antennas*. New York: Wiley Inc., 1998

[4] N. Fourikis, *Phased Array-Based Systems and Applications*. New York: Wiley Inc., 1997

[5] A. Bhattacharyya, *Phased Array Anntennas*. New York: Wiley Inc, 2006

[6] G. Marconi, "Directive antenna," *Proc. Royal Soc.*, London, Vol. 77A, p. 413, 1906.doi:10.1098/rspa.1906.0036

[7] H. H. Beverage, C. W. Rice and E. W. Kellogg, "The wave antenna," *J. Am. Inst. Electr. Eng.*, Vol. XLII, pp. 258–269, March 1923, pp. 372–381, April 1923, pp. 510–519, May 1923, pp. 636–644, June 1923, pp. 728–738, July 1923.

[8] H. Yagi and S. Uda, "Projector of the sharpest beam of electric waves," *Proc. Imp. cad.* (Tokyo) Vol. 2, p. 49, 1926.

[9] H. Friis, "A new directional receiving system," *IRE Proc.* Vol. 13, No. 6, pp. 685–708, Dec. 1925.doi:10.1109/JRPROC.1925.220990

[10] H. T. Friis and C. B. Feldman, "A multiple unit steerable antenna for short-wave reception," *IRE Proc.* Vol. 25, No. 7, pp. 841, July 1937.doi:10.1109/JRPROC.1937.228354

[11] B. T. Neale, "CH—The first operational radar," http://www.radarpages.co.uk/mob/ch/chainhome.htm. Copyright by Dick Barrett

[12] H. T. Friis and W. D. Lewis, "Radar Antennas," *Bell Syst. Tech. J.*, Vol. XXVI, No. 2, pp. 219–317, April 1947.

[13] A. G. Fox, An adjustable wave-guide phase changer, *Proc IRE.*, Vol. 35, pp. 1489–1498, Dec 1947.doi:10.1109/JRPROC.1947.234574

[14] W. H. Kummer, "Feeding and phase scanning" in *Microwave Scanning Antennas*, Vol. III, R. C. Hansen, Ed. New York: Academic Press, p. 82.

[15] C. A. Fowler, "Old radar types never die; they just phased array," *IEEE AES Syst. Mag.*, pp. 24A–24L, Sept. 1998.

[16] R. A. Watson Watt and J. F. Herd, "An instantaneous direct reading radiogoniometer," *J. IEE* (London), Vol. 64, pp. 611–622, 1926.

[17] R. J. Mailloux, *Op Cit.*, chapter 5

[18] B. Munk et al., "A low profile broadband phased array antenna," in *IEEE Antenna and Propagation Society Symposium*, pp. 448–451, 9–12, June 2006, Albuquerque, NM

[19] M. W. Elsallal and D. H. Schaubert, "Electronically scanned arrrays of Dual-Polarized Doubly-Mirrored Balanced Antipodal Vivaldi (DmBAVA) based on modular of elements," in *IEEE Antenna and Propagation Society Symposium*, pp. 887–889, 9–12, June 2006, Albuquerque, NM

[20] C.T. Tai, " The optimum directivity of uniformly spaced broadside arrays of dipoles," *IEEE Trans. AP-12*, pp. 447–454, 1964 .

[21] R. C. Hansen, *Op Cit.*, chapter 1, pp. 89–91.

[22] T. T. Taylor, "Design of Line-Source Antenas for Narrow Beamwidth and Low Side-lobes", *IRE Trans. ntennas Propagation*, pp. 16–28, Jan. 1955.

[23] E. D. Sharp, "A triangular arrangement of planar-array elements that reduces the number needed", *IRE Trans. Antennas Propagation*, pp. 126–129, March 1961.doi:10.1109/TAP.1961.1144967

[24] C. T. Tai, *Op Cit.*

[25] J. Butler and R. Lowe, Beam forming matrix simplifies design of electronically scanned arrays, *Elect. Design*, Vol. 9, pp. 170–173, April 12, 1961.

[26] W. Rotman and R. F. Turner, "Wide angle microwave lens for line source applications," *IEEE Trans. AP-11*, pp. 623–632, 1963

[27] H. Gent, "The bootlace arial," *Royal Radar Establish. J.*, pp. 447–454, Oct. 1957.

[28] D. Archer, "Lens fed multiple beam arrays," *Microwave J.*, pp. 37–42, Oct. 1975

[29] J. Blass, "The multidirectional antenna: a new approach to stacked beams," *1960 IRE International Convention Record*, pp. T.1, 48–50.

[30] S. Stein, "Cross couplings between feed lines for multibeam antennas due to beam overlap", *IEEE Trans. AP-10*, pp. 548–55, Sept. 1962.

[31] J. L. Allen, "A theoretical limitation on the formation of lossless multiple beams in linear arrays," *IEEE Trans. AP-9*, pp. 350–352, July 1961,

[32] J. L. Allen, "Phased array radar studies, July 1960 to July 1961", Technical Report # 236, (u), Lincoln Laboratory, MIT, Nov. 13, 1961, Part 3, chapter 1, DDC271724 measured pattern.

CHAPTER 2

Element Coupling Effects in Array Antennas

2.1 INTRODUCTION

The patterns plotted and the results discussed in Chapter 1 (and the equations used later in Chapter 3) assume that all array elements radiate with the same pattern, and so describe the array behavior in terms of an array factor multiplied by a single element pattern. The presentation in those two chapters is directly useful as long as the basic elements have current or field distributions whose shape does not change as a function of scan or location in the array. We refer to them as single mode elements, and their radiated pattern is that of the isolated element. In this case one must realize that the currents or aperture fields are not necessarily proportional to the applied signals, so it is necessary to include the mutual coupling analysis to obtain the source values to produce the desired currents.

Those equations are also useful using array element patterns that are measured or computed, and that already include the mutual coupling between elements. Each array element pattern in the array will be different, but for a large array with terminated edge elements it may be adequate to use an average array element pattern.

2.1.1 Array Element Patterns

The pattern of an element in an array, or the array element pattern, has been a useful tool for many years. Array element patterns are intuitively simple and easy to measure, but difficult to calculate because they depend upon the mutual interaction (coupling) between elements in the array. The array element pattern for any element is measured by applying an incident signal to that element with all other array elements terminated in matched loads, and measuring the resulting far field received signal at some angular location in space using some probe element. Since Maxwell's equations are linear, the element field patterns of the individual elements are summed to yield the pattern of the whole array. The field pattern is best seen as a term in the scattering matrix that includes all the array elements and the near or far field probe that receives from or alternatively transmits toward the array. The scattering

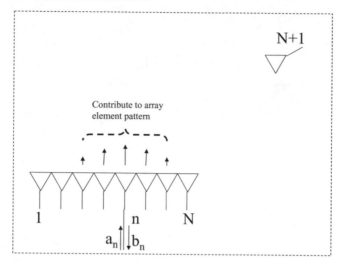

FIGURE 2.1: Microwave network representation of N-element array and far field probe.

matrix representation is especially appropriate since it is a direct measurable, using single mode transmission feed lines of any sort, and input and output voltage signals as the measurable parameters.

In general, antenna elements in an array, especially an array that is scanned, behave almost entirely differently from isolated elements because exciting that single antenna would necessarily induce currents on all of the array elements, which then all radiate. The single input port actually excites the whole array, but most strongly the nearest neighbors. One would then expect, and will usually find, that the element radiates with a narrower pattern than if it were isolated. Figure 2.1 depicts an array with one driven element and indicates neighboring elements radiating to produce the array element pattern of that element. It also depicts this general network, with $N + 1$ elements, an array of N elements and another element in the far field that we label element $N + 1$. The nth element is excited by an incident signal voltage a_n, and the transmission lines applying this signal. Reflected signals b_n eminate from the terminals of each element. All elements are terminated in matched loads. This is then a classic network described by a linear and reciprocal scattering matrix.

The matrix includes all mutual coupling within the array of external scatterers. The signals emanating from each of $N + 1$ terminals are related by the usual matrix:

$$\mathbf{b} = \mathbf{Sa} \qquad (2.1)$$

where the column vector $\mathbf{b}$ are the signals out of each terminal, and the incident signals make up the vector $\mathbf{a}$. These incident signals could include $\mathbf{a}$ phase shift or time delay and a complex weight, and so control the array pattern. Loading all but some chosen element "n" with matched loads and setting all $a_m = 0$ for $m \neq n$, the signals out of every element port become

$$b_m = S_{mn}a_n. \tag{2.2}$$

The location of the probe antenna is some function of θ and ϕ, and r is the distance to the antenna. As θ and ϕ are changed, at constant distance r the received probe signal traces out the element pattern.

$$b_{N+1}(\theta, \phi) = S_{N+1,n}(\theta, \phi)a_n \tag{2.3}$$

Thus, with some normalized signal a_n at the nth terminal and all others terminated in matched loads, the received signal at the far field probe is simply the field pattern of element n, and $S_{N+1,n}(\theta, \phi)$ is the element pattern of element n. Because of reciprocity, for every array element $S_{N+1,n} = S_{n,N+1}$, so the transmit and receive element patterns for any one element of the array are the same, and the transmit and receive gain is the same for any element when measured in this manner.

When signals are applied to all of the array ports, superposition applies and the signal at the far field probe output is:

$$b_{N+1} = \sum_{n=1}^{N} S_{N+1,n}\, a_n \tag{2.4}$$

In a large array, most of the elements are far from an edge, and those near the edge are excited with reduced amplitude for sidelobe control. Therefore, except for the phase center displacement (a factor $\exp(jknd_x \sin \theta)$ for the linear array), all of the central element patterns are nearly the same, and the array element power pattern for some central element is nearly the same as the normalized array gain pattern for the scanned array. The array element pattern shows all consequences of mutual coupling including blindness and the detailed pattern behavior near grating lobes. It is important to note however that all the array element patterns in an array are different, so using only one central element pattern or some average array element pattern to estimate performance is at best an approximation, and then only a good approximation for large arrays. Element pattern measurement however remains the designers' most powerful tool.

2.1.2 Array Element Patterns for Single and Multi-mode Elements

The 'single mode' assumption is a good approximation for elements that have all dimensions on the order of a half wavelength or less. These kinds of elements (dipoles, thin slots, narrow band microstrip patches over very thin substrates, etc.) radiate with a well-defined pattern (the pattern of an isolated element) whether at the center or at the edges of an array.

This means that the results of Chapter 1 are valid for arrays of these "single mode" elements *if one knows the currents or incident voltages to apply to the array factor*. Unfortunately,

Due to mutual coupling between the elements, these currents are not simply proportional to the applied voltages. In terms of the nomenclature of Chapter 1, one needs to force the currents to have the values a_p, but the actual applied voltages are unknown and must be determined through the inversion of a mutual coupling matrix.

Other elements are more complex than these "single mode" elements, and even their fundamental current distributions or aperture fields are different at different elements, especially near the array edges. Examples are rectangular waveguides that support higher order modes at the aperture and dipoles excited by two wire lines that develop strong radiating currents on the feed lines for certain scanning angles. These were principally responsible for severe and unpredicted "blindness" discovered in early radar arrays. If these higher order mode distributions are left out of the calculations, one does not discover the catastrophic array blindness. This fact was clearly presented in the early work of Farrel and Kuhn [1] who compared waveguide arrays calculated with and without higher order modes, and showed that including at least the first odd-symmetry mode was essential for predicting blindness.

Multiple modes can be present in the measured or computed scattering matrix, and as long as all elements and their higher order modes are terminated as they would be in the array on transmit and receive, the summation will lead to the correct radiated pattern.

In summary then, the Chapter 1 equations can be used universally as long as the individual array element patterns are used instead of factoring out a common element pattern. They can also be used with a common pattern (the isolated pattern) if the actual currents are evaluated by including the mutual coupling. Beginning with the work of Steyskal and Herd [2], who measured the array element patterns from a small array and solved for the required excitation to radiate a required pattern, other authors have performed this procedure and demonstrated extremely good results, even for very low sidelobe distributions [3].

2.2   RADIATION OF WIRE AND APERTURE ANTENNAS

2.2.1   The Radiation of a Single Wire or Aperture

Figure 2.2 shows the relevant coordinate systems, and the two of the primary sources of radiation that we will consider; thin wires and apertures over ground screens.

Radiation from current elements without the presence of dielectric material is most often evaluated using the vector magnetic potential $\mathbf{A}$. At any point in space (x, y, z) the potential $\mathbf{A}$ is given in terms of an integral over the electric current density $\mathbf{J}$, enclosed by some surface S as:

$$\mathbf{A}(x, y, z) = \frac{\mu_0}{4\pi} \int_S \mathbf{J}(s') \frac{e^{-jkR}}{R} ds', \quad R = \sqrt{(x - x')^2 + (y - y')^2 + (z - z')^2} \qquad (2.5)$$

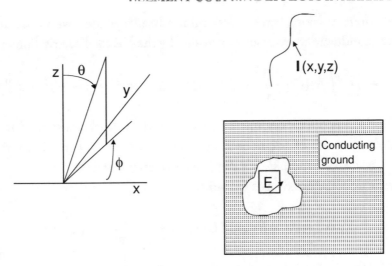

FIGURE 2.2: Coordinate system, wire and aperture antennas.

where the primed coordinates are the indicate the location of the current element, (x, y, z) is the observation point and dv' is the increment of volume.

In the far field, we replace the Green's Function e^{-jkR}/R by the approximation below:

$$\left(\frac{e^{-jkR}}{R}\right) = \left(\frac{e^{-jkR_0}}{R_0}\right)e^{+j\hat{\rho}\cdot\mathbf{r}'}, \qquad \hat{\rho} = \hat{\mathbf{x}}u + \hat{\mathbf{y}}v + \hat{\mathbf{z}}\cos\theta; \quad \mathbf{r}' = \hat{\mathbf{x}}x' + \hat{\mathbf{y}}y' + \hat{\mathbf{z}}z' \qquad (2.6)$$

and where R_0 is the center of the coordinate system and $\mathbf{r}'$ is a vector from the center of the coordinate system to the coordinate of ds' and $\hat{\rho}$ is a unit vector in the direction of the point of observation. For the case of the Magnetic potential above, this yields:

$$\mathbf{A}(u, v) = \frac{\mu}{4\pi}\left(\frac{e^{-jkR_0}}{R_0}\right)\int_s \mathbf{J}(x', y', z')e^{jk(ux'+vy'+z'\cos\theta)}ds' \qquad (2.7)$$

Electric and magnetic fields are related by

$$\mathbf{E}_A = -j\omega\mathbf{A} - \frac{j}{\omega\mu\varepsilon}\nabla(\nabla\cdot\mathbf{A}) \text{ and } \mathbf{B} = \nabla\times\mathbf{A} \qquad (2.8)$$

In the far field these parameters are given approximately by

$$\mathbf{E}_A = -j\omega\mathbf{A_T} \text{ and } \mathbf{H}_A = -\frac{j\omega}{\eta}\hat{\rho}\times\mathbf{A_T} \qquad (2.9)$$

where η is the characteristic impedance of free space, $\eta = \sqrt{\mu/\varepsilon}$, and the subscript implies that it is only necessary to include the components of $\mathbf{A}$ (A_θ and A_ϕ) transverse to the radial $\hat{\rho}$.

Aperture antennas (slots, waveguides) are nearly always used with ground screens and, assuming perfect conductivity are readily expressed by the Vector Electric Potential $\mathbf{F}$ with:

$$\mathbf{F}(x, y, z) = \frac{\varepsilon_0}{2\pi} \int \mathbf{M}(s') \frac{e^{-jkR}}{R} ds', \qquad R = [(x - x')^2 + (y - y')^2 + z^2]^{1/2} \qquad (2.10)$$

In this form the potential assumes a perfectly conducting planar ground screen at $z = 0$, and $\mathbf{M}$, the surface density of magnetization is related to the tangential electric field in the aperture $\mathbf{M} = -\hat{\mathbf{n}} \times \mathbf{E_{AP}}$, where $\mathbf{E_{AP}}$ is the tangential aperture field.

The electric and magnetic fields are written:

$$\mathbf{E_F} = -\frac{1}{\varepsilon} \nabla \times \mathbf{F} \text{ and } \mathbf{H_F} = -j\omega\mathbf{F} - \frac{j}{\omega\varepsilon\mu} \nabla(\nabla \cdot \mathbf{F}) \qquad (2.11)$$

And the far fields are:

$$\mathbf{E_F} = j\omega\eta\hat{\boldsymbol{\rho}} \times \mathbf{F_T} \quad \text{and} \quad \mathbf{H_F} = -j\omega\mathbf{F_T} \qquad (2.12)$$

The two potential functions still assume that the current density $\mathbf{J}$ or aperture field $\mathbf{E}_{ap}$ distributions are known, and so we use the boundary condition that the tangential electric field on the wire is zero ($\hat{\mathbf{n}} \times \mathbf{E} = 0$) where here the unit vector $\hat{\mathbf{n}}$ is normal to the surface enclosing the current, while the boundary condition on any aperture in a ground screen is that the tangential magnetic fields $\mathbf{B}$ and $\mathbf{H}$ are continuous across the aperture in the screen between regions 1 and 2 or $\hat{\mathbf{n}} \times (\mathbf{B}_1 - \mathbf{B}_2) = 0$, where region 1 may be free space bounded by the screen, and region 2 a bounded feed network, like a rectangular waveguide or parallel plate waveguide, and again $\hat{\mathbf{n}}$ is the unit vector normal to space 2 and into space 1. Both electric and magnetic potential functions are used in the presence of additional ground screens by using the appropriate magnetic and electric image current sources [4].

2.2.2   Radiation from a Dipole

Since the (infinitesimally thin) dipole and the slot are related by duality, only the dipole will be considered in this section.

Assume that a single dipole of radius 'a' and length h directed in the $\hat{\mathbf{z}}$ direction is excited by a source of voltage that is a delta function $V_0\delta(z)$ such that the voltage is zero unless $z = 0$, and the integral over the one-dimensional source is V_0. Figure 2.3 shows the dipole with this source. Since the dipole is assumed to be very thin compared to its length, all the current must flow in the z direction and the tangential electric field E_z^s radiated by the dipole must satisfy the boundary condition that the total tangential field must be zero on the surface of the perfectly conducting dipole. Thus E_z^s, the field scattered by the dipole, must be equal to the negative of

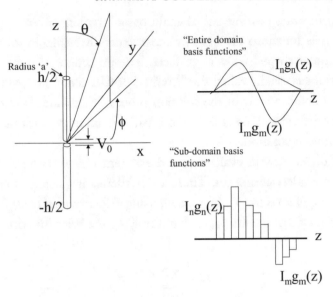

FIGURE 2.3: Dipole antenna and entire domain and sub-domain basis functions.

the applied field $V_0\delta(z)$ given:

$$E_z^s(z) = \frac{-j}{\omega\mu\varepsilon}\left(\frac{\partial^2}{\partial z^2} + k^2\right) A_z(z) = -V_0\delta(z) \qquad -h/2 \leq z' \leq h/2$$

or

$$(2.13)$$

$$E_z^s(z) = \frac{-j}{\omega\varepsilon\mu}\int I(z')\left(\frac{\partial^2}{\partial z^2} + k^2\right)\frac{e^{-jkR}}{4\pi R}dz' = -V_0\delta(z) \qquad -h/2 \leq z' \leq h/2$$

where $R = \sqrt{(z-z')^2 + a^2}$ and "a" is the dipole radius.

The first form of this equation is called the Hallén's equation and often has been used when iterative solutions have been obtained. The second form of this equation is called Pocklington's equation, and is the more commonly used for numerical solutions.

Either form of Eq. (2.13) needs to be satisfied at every point on the dipole surface, and since there is no closed form solution possible, the equation is usually solved by expanding the current in a series of "basis functions" illustrated in Fig. 2.3. The basis functions can be "entire domain" basis functions, wherein each function spans the entire dipole, or "sub-domain" basis functions, wherein the basis function is defined over a short length of the dipole and the combination of many such functions spans the entire length of the wire. The entire domain type of basis function, shown top right is often chosen to be some "complete" set, for example the Fourier Series components that are particularly suited to rectangular apertures. The sub-domain basis functions shown are called pulse functions or piecewise constant functions, but there are a number of other possible sub-domain basis functions given in graduate level texts [4]. Sub-domain basis functions are used by most generally available software and are most appropriate for complex shaped wires or surfaces. The procedure to reduce the integrals to a set of matrix equations is usually called the "Method of Moments" and is widely referenced in the literature. For a detailed view of the solution process, the reader is referred to the text of Harrington [5] that describes some of the various basis function expansions and gives details of the Method of Moments solution.

Either equation leads to the evaluation of the current $I(z)$ consistent with both Maxwell's equations and the boundary condition. There are subtleties neglected in this description and for these the reader is referred to [6]. In general, using either type of basis function, one writes the current as a series of coefficients I_p, each multiplying the basis function $g_p(z)$.

$$I(z) = \sum_{p=1}^{M} I_p g_p(z') \qquad (2.14)$$

After substituting this series approximation for the current into Eq. (2.13), the integral equation is written as:

$$-V_0 \delta(z) = \sum_{p=1}^{M} \int I_p g_p(z') K(z, z') dz' , \quad K = \frac{1}{j\omega\mu\varepsilon} \left(\frac{d^2}{d_z^2} + k_0^2 \right) \frac{e^{-jk_0 R(z,z')}}{R(z, z')} \qquad (2.15)$$

and K is called the Pocklington kernel.

This integral equation is then converted to a set of algebraic equations by taking an inner product of the whole equation with a set of "weighting functions", $w_q(z)$, such that for any $h(z)$ the inner product is defined as the integral over the domain of the weighting function:

$$\langle w_q, h \rangle \equiv \int w_q(z) h(z) dz \qquad (2.16)$$

The weighting functions $w_q(z)$ can in fact be the same set of functions as the basis functions, in which case this special case of the Method of Moments is called Galerkin's method, and has special variational characteristics. Taking this inner product, the rows of the final equation are written

$$-V_0 w_q = \left\langle w_q(z) \sum_{p=1}^{M} I_p \int g(z') dz' \right\rangle \quad \text{for } 1 \le q \le M \text{ and } 1 \le p \le M \qquad (2.17)$$

When integrated this is seen as an $M \times M$ impedance matrix and this form highlights the impedance relationship between the basis current and the applied voltages.

$$V_q = \sum_p Z_{qp} I_p \quad \text{for } 1 \leq p \leq M \text{ and } 1 \leq q \leq M, \qquad (2.18)$$

or simply $V = ZI$, for voltage and current vectors V and Z and square matrix Z.

2.2.3 Radiation from a Dipole in an Array of Terminated Dipoles

The radiation from an array of N wires or apertures is evaluated using the same procedure as that for the single element. The boundary conditions must however be imposed at all N surfaces, and at each of the surfaces the incident field must include the electric field from all the other elements operating at a distance.

The potential functions (Eqs. 2.5 and 2.8) for an array take the form of a summation over all of the radiating structures. For example, an array with dipole elements located at $x = ndx$, must satisfy the Pocklington equations on each dipole.

To solve this set of equations for sub-domain basis functions, one numbers the segments on one antenna from 1 to M, on the next from $M + 1$ to $2M$ and so on until the last segment on the Nth antenna, numbered $M \times N$. After taking the inner product of that equation with the set of functions w_q, one obtains a final equation that has the same form as Eq. (2.17).

$$- V_0 w_q(0) = \left\langle w_q(z) \sum_{p=1}^{M \times N} I_p \int g_p(z') K(x, z, x', z') dz' \right\rangle \quad \text{for}$$
$$1 \leq q \leq M \times N \text{ and } 1 \leq p \leq M \times N \qquad (2.19)$$

with R written as:

$$R(x, z) = R(ndx, z) = \sqrt{(z_p - z')^2 + (ndx - x')^2 + a^2},$$

where here the very small term a^2 has been included to avoid the singularity at $x' = ndx$ when $z' = z_p$. The resulting set of equations has the same form as Eq. (2.16), but now includes all of the elements.

$$V_q = \sum Z_{qp} I_p \quad \text{for } 1 \leq q, p \leq M \times N \qquad (2.20)$$

or in matrix form:

$$V = ZI \qquad (2.21)$$

with V and I column vectors, and the Z matrix coefficient

$$Z_{qp} = \int\limits_{z_q-1}^{z_q} w_q(z) \int\limits_{z_p-1}^{z_p} g_p(z')K(z,\,z')dz'dz \qquad (2.22)$$

If any antenna port is simply terminated in a load the procedure is to replace the port with the voltage drop across the terminating resistor, and consider that a voltage source, so in the special case of an array with only one element excited, and the rest loaded with an impedance Z_{mm}, the self impedance of the dipoles, then each row of the impedance matrix sees an equivalent source $V_m = -Z_{mm}I_m$. This set of equations can then be solved for the current coefficients I_p and the radiated fields.

Although the expressions above are used to produce higher order results using any number 'M' of basis functions, many of the basic principles can be illustrated with a single mode approximation to the current. The following results are obtained using such an approximation. The current on any nth element is assumed:

$$I(z) = I_0 \sin\left[k\left(\frac{h}{2} - |z|\right)\right]. \qquad (2.23)$$

This expression is exact for half wave elements that are infinitesimally thin, and is a fair approximation for elements less than a wavelength long [4]. The terms of the impedance matrix are given in several texts [4, 7], and so will not be reproduced here. This simple lowest order approximation will be used to compute the data for the next section.

Figure 2.4 shows the radiated patterns for small arrays with a single central element excited and all others terminated. Figure 2.4(a) shows the azimuth radiation pattern of a single dipole, and that of the dipole when numbers of additional passive dipoles are added with the original one centered. The passive dipoles are not excited, but loaded with an impedance $ZL = 73 + j42.5\ \Omega$, the radiation resistance of the single isolated dipole. A set of N simultaneous equations replace the above.

The patterns in Fig. 2.4 (a–d) are the element patterns of the central element of an array of 1, 7, 21, 41 elements etc. These tend progressively toward an asymptotic infinite array pattern limit as the number of elements is increased. Data for an array of 101 elements are essentially the same as that for 41 elements, and so are not shown. It is important to note here that the coupled current on the added antennas significantly distorts the radiated pattern of the isolated antenna, and that the element pattern is well formed with only 21 elements in the array. It is essentially the same for arrays with 21 elements or an infinite array of elements.

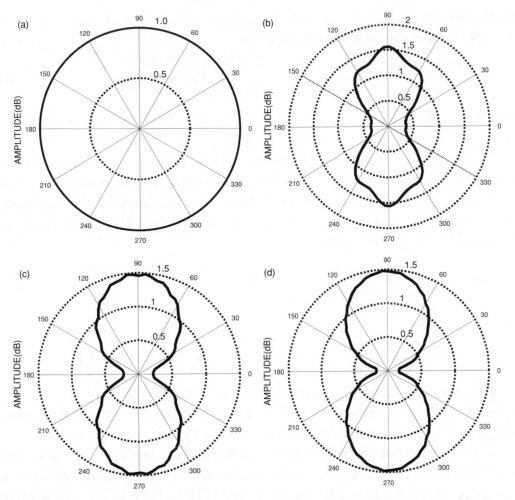

FIGURE 2.4: Array element (azimuth) patterns for arrays of N-elements. (a) $N = 1$, (b) $N = 7$, (c) $N = 21$, and (d) $N = 41$.

2.3   ARRAY ELEMENT PATTERNS IN AN INFINITE ARRAY

The figures shown in the last section indicate that the element pattern of an isolated radiator is very different from the same element in an array. The array element pattern in an infinite array of scanned elements has been shown to be far more descriptive of actual array behavior in all but relatively small arrays, because it displays, in one calculation, the entire scan behavior of the scanning array at all angles and all polarizations. In addition, the infinite array calculation is far simpler than the evaluation of radiation from a finite array.

For example, in an infinite two-dimensional array with a rectangular grid, the potential function of Eq. (2.5) takes on a special form when the number of elements becomes infinite, for the kernel summed over in infinite number of $\exp(-jkR_{mn})/R_{mn}$ is transformed using the Poisson Summation Formula. The initial form is:

$$\mathbf{A}(x, y, z) = \frac{\mu_0}{4\pi} \int \sum_{m=-\infty}^{m=\infty} \sum_{n=-\infty}^{n=\infty} \mathbf{J}_{mn} \frac{e^{-jkR_{mn}}}{R_{mn}} dv', \qquad (2.24)$$

where $R_{mn} = \sqrt{(x - md_x - x')^2 + (y - nd_y - y')^2 + (z - z')^2}$ is the distance between any point (x, y, z) and a point on the element m, n at (x', y', z').

The current $\mathbf{J}_{mn}$ is assumed to be periodic and have the exponential relationship to the current at the zero element as $\mathbf{J}_{mn} = \mathbf{J}_{00} \exp(-jk(u_0 md_x + v_0 nd_y))$, since this is the phase or time delay progression to scan the beam to the angles (u_0, v_0). Inserting this expression for the summation above, and applying the Poisson Summation Formula to the resulting double summation leads to the expression below

$$\mathbf{A}(x, y, z) = \frac{-j2\pi}{d_x d_y} \frac{\mu_0}{4\pi} \sum_{p=-\infty}^{p=\infty} \sum_{q=-\infty}^{q=\infty} \int \frac{e^{-jk[u_p(x-x')+v_q(y-y')]-jK_{pq}|z-z'|}}{K_{pq}} \mathbf{J}_{00} dv' \qquad (2.25)$$

where

$$u_p = u_0 + \frac{p\lambda}{d_x}, \quad v_q = v_0 + \frac{q\lambda}{d_y}, \quad \text{and } K_{pq} = k\left[1 - u_p^2 - v_q^2\right].$$

This expression shows the summation over the infinite number of individual sources of spherical waves, now summed to an infinite number of waves with propagation constants $k_x = k(u_0 + (p\lambda/d_x))$ in the x direction and $k_y = k(v_0 + (q\lambda/d_y))$ in the y direction.

With the summation written in this form, the potential function is seen to consist of waves in the directions of the grating lobes. Most of these waves are in imaginary space however, and so do not actually radiate, but they do all need to be included in the summation until their contribution becomes negligible. The series is rapidly converging however and so a more efficient form to evaluate.

The per-element gain for a unit cell of the infinite two-dimensional array is given

$$G_e = \frac{4\pi A_{\text{cell}}}{\lambda^2} (1 - |\Gamma|^2) \cos \theta \qquad (2.26)$$

with Γ the element reflection coefficient. This expression is readily derived for the infinite array case because all the radiation occurs from one term of the grating lobe series at the beam scan direction (u_o, v_o) and the power lost is the reflected $(1 - |\Gamma|^2)$. The $\cos \theta$ is incurred due to the array projection in the θ direction. This familiar expression is used as the array element power pattern for elements in a large array.

Most of the early studies of infinite two-dimensional scanning arrays used dipole or thin slot elements, usually with the dipoles over a conducting screen. These studies described the

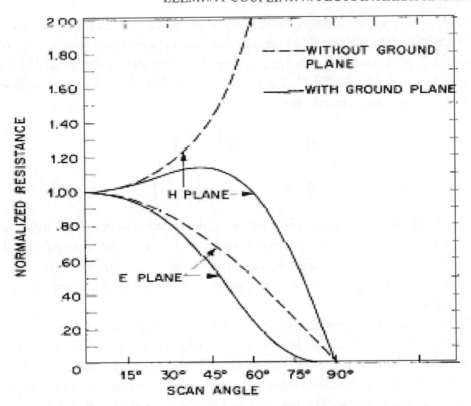

FIGURE 2.5: Normalized input resistance of array of dipoles with and without ground plane. (After Edelberg and Oliner [11] with permission).

scan properties of idealized elements and often focused on the input parameters of driving point resistance or reactance. The publications by Wheeler [8], Carter [9], Stark [10], and Edelberg and Oliner [11] are examples of these calculations. Figure 2.5 is from the paper by Edelberg and Oliner and shows the normalized resistance of a dipole element in a two-dimensional infinite array with and without a ground screen below the array. Clearly there is a significant change in resistance as the array is scanned.

In 1969, Wheeler [12] conducted a study that contributed much insight into array behavior by analyzing the radiation of a continuous current sheet in transmit and receive conditions. The electric current sheet approximates, the behavior of a sheet of closely spaced electric dipoles, while the magnetic current sheet behaves similar to a metallic sheet of slots. Wheeler showed that the current sheet model presented a reflection coefficient of a particularly simple form. For an electric current sheet transmitting a wave at an angle θ in the H plane, Wheeler showed that the resulting variation in element resistance with angle has the normalized value, $R/R_0 = 1/\cos\theta$, while the array scanned in the E plane has the inverse $R/R_0 = \cos\theta$.

Alternatively, for an array represented by a magnetic current sheet, the variation in conductance is similar, with the H plane scan resulting in normalized conductance $G/G_0 = \cos\theta$, and for E-plane scan $G/G_0 = 1/\cos\theta$. In each of these cases the normalizing resistance is that of free space, namely 377 Ω. These result in the following reflection coefficients for either electric or magnetic current sheets:

$$\text{H-scan:} \quad \Gamma = \left(\tan\frac{\theta}{2}\right)^2$$

$$\text{E-scan:} \quad \Gamma = -\left(\tan\frac{\theta}{2}\right)^2 \qquad (2.27)$$

The sketch in Fig. 2.6 shows this coefficient for the transmit case. This figure indicates that the array reflection varies over a wide range with scan. For the conducting sheet, although the magnitude of the reflection coefficient is the same for both polarizations, the phase is actually opposite. This behavior precludes obtaining good scan control with any simple impedance matching circuit.

These relationships, though derived for continuous surfaces, do give insights to the periodic array behavior for angles less than the grating lobe angle. They give no account of grating lobe effects, or the particulars of arrays and feeds that result in the "blindness" phenomenon that will be described later, and they certainly do not account for higher order mode effects on the elements. The early slot and dipole array studies [8–12] did account for the grating lobe onset, but also analyzed such simple structures (thin dipoles and slots) that they did not account for any of the higher order effects that led to the phenomenon of array blindness. In 1985 Pozar [13] extended Wheeler's current sheet approach to consider periodic arrays of printed elements over dielectric substrates with and without ground planes, but using a current sheet model with a few additional Floquet modes (grating lobe series terms). His primary conclusion was that for simple elements over dielectric substrates, lattice effects, not element type, are more important for prediction of the location of blindness in arrays or printed antennas. This conclusion cannot be generalized to all arrays, and the next paragraph details cases wherein both the lattice and the element type combined to cause the blindness condition (Fig. 2.6).

Figure 2.7 shows data from the infinite array study published by Farrel and Kuhn [1], demonstrating array "blindness". This phenomenon results in such a severe mismatch at the array face, as to present nearly 100% reflection. The presented data are for an array of waveguide apertures, but such blindness has been found to exist in dipole arrays, microstrip arrays and wide band arrays like Vivaldi structures. In the waveguide array the blindness is caused by the radiation from the first odd mode in the feed waveguides, in dipole arrays it is caused by radiation from the dipole feed lines [14], in microstrip arrays it is primarily caused by the wave

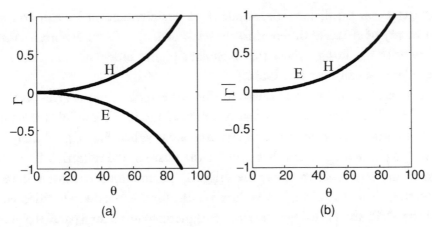

FIGURE 2.6: The variation of reflection with angle, in the idealized model. (a) For the case of electric dipoles, from the sending viewpoint. (b) For either kind of dipoles, from either viewpoint (After Wheeler [11], with permission).

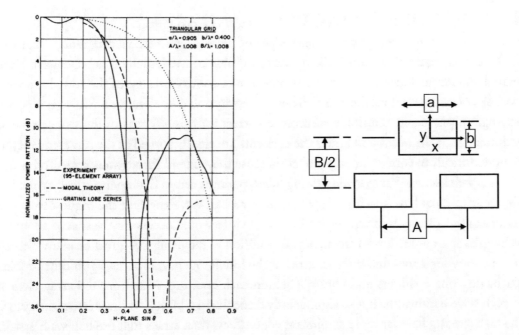

FIGURE 2.7: Array element pattern illustrating blindness (After Farrell, Jr. and Kuhn [1], with permission).

that can propagate on the dielectric substrate or superstrate, and in Vivaldi arrays it can be caused by a variety of effects at different frequencies, including feed network resonances [15]. These results were the first to show that including higher order modes in an analysis was essential in order to reveal array blindness.

An early explanation of the behavior for the waveguide case was given in the paper by Knittel et al. [16]. The phenomenon is that of a normal mode of a modified structure, one in which the feed ports are shorted or reactively terminated. When the original array happens to be excited with a phase progression that matches the phase of the reflection from the reactive termination in the modified structure, the original and modified structures have the same solution, and there is no radiation, just as there is none for the reactively terminated structure. A later study by Mailloux [17] using a full electromagnetic solution for a parallel plate array with fences, explored both the surface wave on a reactively terminated surface and the radiating array solution and showed that at the blindness angle the two structures were electromagnetically identical.

2.4 FINITE ARRAYS AND WIDE-BAND ARRAYS

The elements and element patterns near edges of finite arrays behave differently than those near the center, that behave much like those in infinite arrays. Usually this behavior is simply tolerated and not included in the array manufacture or control because of the added expense of providing corrections within the array. Where necessary and cost effective however it is possible to reprogram phase and amplitude controls to correct for this effect as in the work previously mentioned [2, 3]. In wide band arrays the elements are closely spaced at the low frequency and it is more difficult to correct for edge effects because they extend throughout more of the array.

It is well known that to avoid grating lobes, periodic arrays that operate over a wide band of frequencies must have spacing of approximately a half wavelength at the highest frequency. This means that elements in an array of 10:1 bandwidth would necessarily be spaced about 0.05 wavelengths apart at the lowest frequency. In addition to requiring 100 times as many elements than necessary for a two-dimensional array at the low frequency, it poses significant problems in packaging phase shifters and controls behind each aperture. Matching the array elements over such wide bandwidths is also significantly difficult. In addition to these issues, arrays with such small spacing have far more significant edge effects than arrays that use half wave spacing. Several new references to studies of this effect include those of Boryssenko and Schaubert [18], Hansen [19], and Munk [20].

REFERENCES

[1] G. F. Farrell, Jr. and D. H. Kuhn, "Mutual coupling effects in infinite planar arrays of rectangular waveguide horns," *IEEE Trans.*, Vol. AP-16, pp 405–414, 1968.

[2] H. Steyskal and J. S. Herd, "Mutual coupling compensation in small array antennas", *IEEE Trans.*, Vol. AP-38, No. 12, pp. 1971–1975, Dec 1990.

[3] L. Pettersson and M. Danestig, Ulf Sjöström, *IEEE International Symposium on Phased Array Systems and Technology*, 15–18 Oct. 1996, IEEE CT. No. 96 7H8175.

[4] C. A. Balanis, *Antenna Theory: Analysis and Design*. New York: John Wiley & Sons, Inc., 1997.

[5] R.F. Harrington, *Field Computation by Moment Methods*. New York: Macmillan, 1968.

[6] G. Fikioris andT. T. Wu, "On the application of numerical methods to Hallen's Equation," *IEEE Trans.*, Vol. AP- 49, No. 3, pp. 383–392, March 2001.

[7] H. E. King, "Mutual impedance of unequal length antennas in echelon," *IRE Trans.*, Vol. AP-5, pp. 306–313, July 1957.

[8] H.A. Wheeler, " The radiation resistance of an antenna in an infinite array or waveguide, *IEEE Proc.*, Vol. 16, April 1948, pp. 478–487.

[9] P. S. Carter, Jr., "Mutual impedance effects in large beam scanning arrays," *IRE Trans. Antennas Propagation*, Vol. AP-8, pp. 276–285, May 1960. doi:10.1109/TAP.1960.1144839

[10] L. Stark, "Radiation impedance of a dipole in an infinite planar phased array," *Radio Sci.*, Vol. 1 (New Series) No. 3, March 1966.

[11] S. Edelberg and A. A. Oliner, "Mutual coupling effects in large antenna arrays 2: Compensation effects," *IEEE Trans.*, Vol. AP-8, No. 4, pp. 360–367, July 1960, Part 1 and 2.

[12] H. A. Wheeler, "Simple relations derived from a phased-array made of an infinite current sheet," *IEEE Trans.*, Vol. AP-13, No. 4, pp. 506–514, July 1965.

[13] D. M. Pozar, "General relations for a phased array of printed antennas derived from infinite current sheets", *IEEE Trans.*, Vol. P-33, No. 5, pp. 498–504, May 1985.

[14] J. J. Schuss, "Numerical design of patch radiator arrays," *Electromagnetics*, Vol. 11, pp. 47–68, Jan. 1991.doi:10.1080/02726349108908263

[15] D. M. Pozar and D. H. Schaubert, "Scan blindness in infinite phased arrays of printed dipoles", *IEEE Trans.*, Vol. AP-32, No. 6, pp. 600–610, June 1984.

[16] G. H. Knittel, A. Hessel and A. A. Oliner, "Element pattern nulls in phased arrays and their relation to guided waves," *Proc. IEEE*, Vol. 56, No. 11, pp. 1822–1836, November 1968.

[17] R. J. Mailloux, "Surface waves and anomalous wave radiation nulls on phased arrays of TEM waveguides with fences," *IEEE Trans.*, Vol. AP-20, No. 2, pp. 160–166, March 1972.

[18] A. O. Boryssenko and D. H. Schaubert, "Predicted performance of small arrays of dielectric-free tapered slot antennas," *2001 Antenna Application Symposium* Digest, Monticello, IL, pp. 250–279, Sept. 2001.

[19]   R. C. Hansen, "Anamalous edge effects in finite arrays", *IEEE Trans.*, Vol. AP-47, No. 3, pp. 549–554, March 1999.

[20]   B. A. Munk, *Finite Antenna Arrays and FSS*. IEEE Press, John Wiley and Sons, 2003.

CHAPTER 3

Array Pattern Synthesis

3.1   INTRODUCTION

Array antennas have the advantage of providing detailed control of amplitude and phase across the aperture. This added control can be used to synthesize radiation patterns that have desirable features like low sidelobes, special shaped patterns that suit a particular application, or patterns with controlled nulls and defined sidelobe regions for jammer or clutter suppression.

Pattern synthesis is the subject of this chapter, and the presentation is limited to describing a few of the classic techniques for synthesis. The methods presented here by no means complete the list of useful methods, but they do form the basis for providing for many of the synthesis needs. In addition to these procedures, modern numerical methods are now commonly used for specific applications, but those methods are too numerous to summarize here. I have included one iterative technique, the method of alternating projections that I have found very useful and easy to apply.

The methods provided are all synthesized for the broadside radiation of one-dimensional arrays. Separable two-dimensional distributions for rectangular shaped arrays can be developed as the product of these, but the unique, non-separable distributions for circular arrays have been omitted. Although only broadside radiation is discussed, this is no restriction the (u, v) patterns are used throughout, and it is only necessary to replace u and v by $u - u_0$ and $v - v_0$ to scan the patterns to any desired direction (u_0, v_0).

The synthesis shown here is based on the array factor. Element pattern variation can be readily included by modifying the desired far field to account for the element patterns. This is done by solving for the synthesized aperture parameters (currents or element pattern amplitude, etc.) and then inverting a mutual coupling matrix to determine the excitation necessary. Examples of this are given in the references [1–3].

3.2   SYNTHESIS TECHNIQUES FOR SECTOR PATTERNS

As noted previously, an array of length Nd_x can produce a pencil beam of width approximately $0.866\lambda / Nd_x$ if excited with a uniform illumination. There are times however when it is desirable to synthesize a much wider pattern; one that may be many times the width of the fundamental p-sinc pattern. These are called sector patterns, and are useful for illuminating a larger angular sector with a lower gain illumination. Among the useful sector patterns are the $\csc^2$ coverage for detection of airborne targets and constant gain (pulse shaped) patterns for flooding a relatively larger area for communication and search radar.

Two methods, the Fourier Series and Woodward synthesis techniques are given below to synthesize sector patterns. They are based on orthogonal basis functions. These approximations, though truncated, can be highly accurate, so both of the techniques can be used to synthesize pencil beam patterns, although they are not efficient solutions in the pencil beam case.

3.2.1   The Fourier Synthesis Method

The one-dimensional antenna far field array factor $F(u)$ is evidently a finite Fourier Series, and is periodic in u-space with the interval of the grating lobe separation λ/d_x.

$$F(u) = \sum_n a_n\, e^{jkund_x} \quad \text{for } -\frac{N-1}{2} \le n \le \frac{N-1}{2}\,;\ k = \frac{2\pi}{\lambda}$$
$$\text{and } N \text{ even } (n = \pm 1/2,\ \pm 3/2, \ldots \quad \text{or odd } (n = 0, \pm 1,\ \pm 2 \ldots)$$

(3.1)

From orthogonality, the coefficients a_n are

$$a_n = \frac{d_x}{\lambda} \int_{-(\lambda/2d_x)}^{\lambda/2d_x} F(u) e^{-j\frac{2\pi}{\lambda} nud_x}\, \mathrm{d}u$$

(3.2)

This method provides the best mean squared approximation to the field $F(u)$ for $d_x/\lambda \ge 0.5$, in the absence of grating lobes. For $d_x/\lambda \le 0.5$ some of the pattern is in imaginary space and the solution is not unique. Figure 3.1 shows the results of applying the Fourier Series method with a 64 element array of isotropic radiators to synthesize a pulse type pattern with unity 1 over the range $-.4 \le u \le .4$ and zero otherwise. The pulse pattern is shown dashed on the figure. The synthesis produces a pattern with very low ripple within the pass-band, and about -21 dB sidelobes.

3.2.2   The Woodward Method

P. M. Woodward [4, 5] developed a technique based on the use of the p-sinc functions of a linear array as basis functions for the synthesis of array patterns.

The p-sinc patterns of uniformly illuminated arrays with scan angles $u_i = (\lambda/Nd_x) i$ have orthogonality properties over the interval $-1 \le u \le 1$ for half wave spacing.

These patterns are written:

$$f_i(u) = \frac{1}{N} \sum_{n=-(N-1)/2}^{n=(N-1)/2} e^{jknd_x(u-u_i)} = \frac{\sin\left[(N\pi d_x/\lambda)(u-u_i)\right]}{N\sin\left[(\pi d_x/\lambda)(u-u_i)\right]}$$

(3.3)

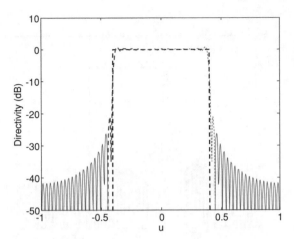

FIGURE 3.1: Fourier series synthesis with 64 element array.

The orthogonality property can be demonstrated by taking the series form of the expression above, multiplying two forms $f_i(u)$ $f_j(u)$, and integrating over the region from −1 to 1 in u-space. The resulting integrals will exhibit the orthogonality, being zero for $i \neq j$. This property is useful since this synthesis can be implemented with constrained lossless multiple beam Butler Matrix networks.

Beyond that, the functions $f_i(u)$ will have a maximum at u_i, while all other $u_j \neq u_i$ will have zeros at the u_i peak. Therefore a function $F(u)$ to be synthesized is only sampled at all u_i, then the pattern can be approximated by summing the set of basis function patterns $f_i(u)$ which we can call the 'constituent beams' since their sum produces the synthesized pattern. The pattern $F(u)$ is thus written:

$$F(u) = \sum_i A_i f_i(u) \quad \text{with} \quad A_i = F(u_i) \tag{3.4}$$

The A_i are the weighting functions multiplying the constituent beams, and by comparing Eq. (3.4) with Eq. (3.1) and using Eq. (3.3) one can show that he actual excitation a_n is:

$$a_n = \sum_i A_i e^{-jku_i d_x n}. \tag{3.5}$$

Figure 3.2 illustrates the use of the Woodward procedure to synthesize the same pulse shape as used for the Fourier method, using the same array. Again the pulse pattern is shown dashed on the figure. In this example the Woodward method showed higher sidelobes and more ripple within the pass-band than the Fourier series method, but otherwise the pattern is a good approximation to the pulse.

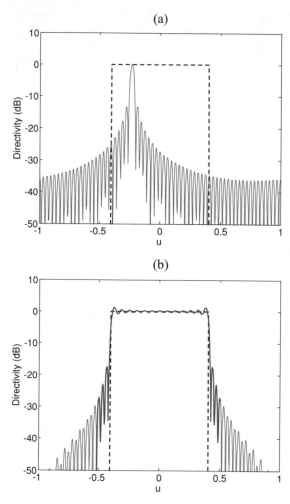

FIGURE 3.2: Woodward synthesis with 64 element array. (a) Basic sinc basis function for $i = -13$ and (b) Synthesized pattern.

3.3 SYNTHESIS OF PENCIL BEAM PATTERNS

Pencil beam patterns are narrow beams intended to illuminate a smaller region than the sector beams. Used in radar and communication because of their higher gain and resolution, the primary parameters of the synthesis are the beamwidth and sidelobe structure.

Earlier methods of pencil beam synthesis were based on creating aperture distributions that were smooth and tapered to lower values at the edges of the aperture to produce lower sidelobes. Among these are a variety of cosine and cosine-squared on a pedestal distributions that are convenient but inefficient. The Dolph–Chebyshev method was one of the first and certainly the most important early method that was a unique mathematical procedure to exactly synthesize a desirable low sidelobe pattern with an array. Details of the procedure are included in many texts and papers [6, 7] and so will not be detailed here, but the technique equates the pattern polynomial for an N-element array to that of a Chebyshev polynomial of order N. The resulting pattern has a constant sidelobe pattern and has the narrowest beamwidth consistant with the chosen sidelobe height. The one disadvantage of the Chebyshev patterns is that the sidelobe height isconstant throughout real space. Thus, a small array with a Chebyshev pattern has a narrow beam and reasonable gain, but as the array is made larger there is more and more energy in the sidelobe structure until increasing the array length begins to produce a pattern with decreased gain. Finally, for large arrays the pattern gain is significantly lower than that of an array with uniform illumination. The only way to combat that occurrence using Chebyshev patterns is to force the sidelobe level to be lower as the array is made larger, and ultimately that becomes an inefficient solution to the practical problem. The Taylor and Bayliss synthesis methods described below present a solution to this problem.

3.3.1   Taylor Line Source Synthesis

In a classic paper, T.T. Taylor [8, 9] developed a synthesis technique that had all the advantages and none of the disadvantages of Dolph–Chebyshev synthesis. Taylor did his analysis on a continuous line source instead of an array, and showed that the far sidelobes are a function only of the line source edge illumination. Further, he showed that if the edge taper has zero derivative, then the far sidelobes decay like $1/n$, where n is the sidelobe index from broadside. The $1/n$ decay is the same as that for a uniform illumination, and so is indicative of high efficiency even for a large array. Larger slopes at the aperture edge provide yet stronger decay of the far sidelobes, but have significantly lower efficiency, so the Taylor synthesis is an excellent compromise that has low sidelobes and high efficiency. In this section we will describe the linear array synthesis, but Taylor later extended his method to circular apertures (Eq. 3.3).

Noting that the zeros of the sinc pattern were equally spaced at $u = nL/\lambda$ for line source length L, Taylor showed that this was also dictated by the edge conditions. He then chose to construct a pattern function that had the same as low or lower than the Chebyshev pattern for the first $\bar{n}$ pattern nulls, and then all nulls at locations $u \geq \bar{n}$ would tend toward the integer locations. This made the first $\bar{n} - 1$ sidelobes at or below the height of the Chebyshev pattern sidelobes, while all others were forced down to follow those of the (uniform illumination) sinc pattern.

The synthesized pattern, normalized to unity is:

$$F(z, A, \bar{n}) = \frac{\sin(\pi z)}{\pi z} \prod_{n=1}^{\bar{n}-1} \frac{1 - (z^2/z_n^2)}{1 - (z^2/n^2)} \qquad (3.6)$$

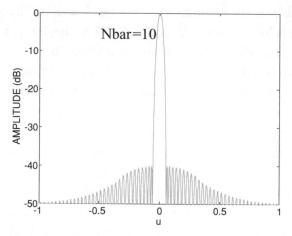

FIGURE 3.3: Taylor synthesis of -40 dB sum pattern using 64 elements and $\bar{n} = 10$.

for $z = uL/\lambda$.

The z_n are the zeros of the synthesized pattern and given by:

$$z_n = \pm\sigma(A^2 + (n - 1/2)^2)^{\frac{1}{2}} \quad \text{for } 1 \leq n \leq \bar{n}$$
$$= \pm n \quad \text{for } \bar{n} \leq n \leq \infty \tag{3.7}$$

where

$$\sigma = \frac{\bar{n}}{[A^2 + (\bar{n} - 1/2)^2]^{1/2}} \tag{3.8}$$

so that at $n = \bar{n}$, $z_n = \bar{n}$ and A is defined by the sidelobe level r, using $A = (1/\pi)\cosh^{-1} r$, for voltage sidelobe level r, the ratio of main beam to sidelobe level, so that $SL_{dB} = 20 \log_{10}(r)$ is a positive number.

The term σ in the above is called the dilation factor, and it stretches out the location of the first $\bar{n}$ pattern zeros, so that they transition from the Chebyshev located zeros to those of the p-sinc function.

The current distribution required to produce this pattern is obtained using the Fourier Series method as:

$$g(x) = F(0, A, \bar{n}) + 2\sum_{m=1}^{\bar{n}-1} F(m, A, \bar{n}) \cos\left(\frac{2\pi mx}{L}\right) \quad \text{for } -L/2 \leq x \leq L/2 \tag{3.9}$$

and the array coefficients are:

$$F(m, A, \bar{n}) = \frac{[(\bar{n} - 1)!]^2}{(\bar{n} - 1 + m)!(\bar{n} - 1 - m)!} \prod_{n=1}^{\bar{n}-1} [1 - m^2/z_n^2] \tag{3.10}$$

The efficiency of this distribution is given by Hansen [10] as:

$$\eta = \frac{1}{1 + 2\sum_{m=1}^{\bar{n}-1} F^2(m, A, \bar{n})} \tag{3.11}$$

The parameter $\bar{n}$ is chosen as a compromise between maximum efficiency and maintaining a monotonic illumination of the aperture. A very small value of $\bar{n}$ results in a pattern that is close to the sinc function, and so has high efficiency, while a large value of the parameter constrains many zeros of the pattern to be close to the Chebyshev pattern and so has low efficiency and for a large array increased illumination near the aperture edges. The table below, from Hansen [10] illustrates this relationship.

TABLE 3.1: Taper efficiency for Taylor patterns

| SL$_{db}$ | MAX η VALUES | | MONOTONIC $\bar{n}$ | |
	$\bar{n}$	η	$\bar{n}$	η
20	6	.9667	3	.9535
25	12	.9252	5	.9105
30	23	.8787	7	.8619
35	44	.8326	9	.8151
40	81	.7899	11	.7729

In most cases the array pattern produced by sampling the line source distribution is an excellent approximation as long as the line source distribution is sampled one-half element spacing from each edge and then uniformly across the array. There have been iterative techniques developed [11, 12] to adjust the sidelobes of the sampled line source pattern, but these are seldom needed except for small arrays.

3.3.2 Bayliss Line Source Synthesis

Bayliss [13], following Taylor, devised a procedure for synthesizing difference patterns with similar characteristics to the Taylor patterns. Using Taylor's notation, Bayliss's procedure constrained the first $\bar{n}$ pattern zeros and thus produced $\bar{n} - 1$ nearly constant sidelobes.

The synthesized pattern is given by:

$$F(z) = \pi z \cos(\pi z) \frac{\prod\limits_{n=1}^{\bar{n}-1} \{1 - (z/\sigma z_n)^2\}}{\prod\limits_{n=0}^{\bar{n}-1} \{1 - [z/n+1/2]^2\}} \tag{3.12}$$

for

$$z = uL/A, \quad \sigma = \frac{\bar{n}+1/2}{z_{\bar{n}}} \quad \text{and } z_{\bar{n}} = (A + \bar{n}^2)^{1/2}$$

The Fourier coefficients B are:

$$g(x) = \sum_{n=0}^{\bar{n}-1} B_n \sin[(2\pi x/L)(n+1/2)] \quad -L/2 \leq x \leq L/2 \tag{3.13}$$

$$B_m = \begin{cases} \dfrac{1}{2j}(-1)^m(m+1/2)^2 \dfrac{\prod\limits_{n=1}^{\bar{n}-1}\left\{1 - \dfrac{[m+1/2]^2}{[\sigma z_n]^2}\right\}}{\prod\limits_{\substack{n=0 \\ n \neq m}}^{\bar{n}-1}\left\{1 - \dfrac{[m+1/2]^2}{[n+1/2]^2}\right\}} & for\ m = 0,1,2\ldots\bar{n}-1 \\[20pt] 0 & for\ m \geq \bar{n} \end{cases} \tag{3.14}$$

The required null locations are given by:

$$z_n = \begin{matrix} 0 & n = 0 \\ \pm\Omega_n & n = 1,2,3,4 \\ \pm(A^2+n^2)^{1/2} & n = 5,6\ldots \end{matrix} \tag{3.15}$$

The coefficients A and Ω are required to construct the patterns, and are tabulated by Bayliss as a function of the sidelobe level SL_{db}. They are not available in closed form, so Baylis presented a table of coefficients for a fourth order polynomial to evaluate A and the Ω_n parameters as a function of the sidelobe level SL_{db}. The polynomial coefficients C_n are given in the table [3.2] below, and reconstructed by the equation:

$$\text{Polynomial} = \sum_{n=0}^{4} C_n[-SL_{dB}]^n \tag{3.16}$$

Table [3.3] lists the coefficients A and Ω corresponding to various sidelobe levels fro 15 to 40 dB below the main beam

TABLE 3.2: Polynomial coefficients to evaluate A and Ω

POLYNOMIAL	C0	C1	C2	C3	C4
A	0.30387530	−0.05042922	−0.00027989	−0.00000343	−0.00000002
Ω_1	0.98583020	−0.03338850	0.00014064	0.00000190	0.00000001
Ω_2	2.00337487	−0.01141548	0.00041590	0.00000373	0.00000001
Ω_3	3.00636321	−0.00683394	0.00029281	0.00000161	0.00000000
Ω_4	4.00518423	−0.00501795	0.00021735	0.00000088	0.00000000

TABLE 3.3: Coefficients A and Ω for specific sidelobe levels

POLYNOMIAL	15	20	25	30	35	40
A	1.0079	1.2247	1.4355	1.6413	1.8431	2.0415
Ω_1	1.5124	1.6962	1.8826	2.0708	2.2602	2.4504
Ω_2	2.2561	2.3698	2.4943	2.6275	2.7675	2.9123
Ω_3	3.1693	3.2473	3.3351	3.4314	3.5352	3.6452
Ω_4	4.1264	4.1854	4.2527	4.3276	4.4093	4.4973

With these coefficients one is able to construct the proper currents and the pattern functions. Figure 3.4 shows a Baylis pattern with sidelobes 40 dB below the main beam and $\bar{n} = 10$.

3.4 NUMERICAL TECHNIQUES FOR SYNTHESIS

The synthesis techniques presented above are direct methods using existing equations for synthesizing a specific desired pattern. These methods are simple to use, but they are constrained to periodic linear or planar arrays. They are also constrained by the use of Fourier series or sinc basis functions for sector patterns and, as in the case of the Taylor and Bayliss functions by their specific design to produce special sidelobe distributions. Numerical methods have been developed to synthesize more general patterns, and some references to these methods are given here [14, 15]. The one technique described below is especially simple to use and very flexible.

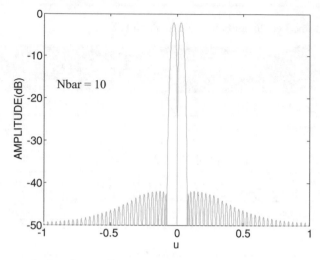

FIGURE 3.4: Bayliss synthesis of -40 dB sum pattern using 64 elements and $\bar{n} = 10$.

3.4.1   The Alternating Projection Method

The technique called "Alternating Projection" or the 'Intersection Approach" is an extremely powerful and versatile procedure for synthesizing the excitation of very general antenna structures. The technique has been used for conformal arrays, planar arrays, for reflector feed synthesis, and I have used it to synthesize a feed distribution for a multiple beam feed for a wideband lens array [16]. However, for the purposes of this section, the details below follow the periodic array and the description of Bucci et al. [17].

The procedure is based on a statement of the fundamental task of synthesis (here reduced to the linear periodic array case), that the radiated patterns for a specific array can be grouped into two sets. One set (B) is the set of all possible patterns that can be radiated by the array. The task of this synthesis is to choose a subset Bc of this set that satisfies the desired result. The resulting solution must be a pattern that is in both sets, i.e., the union or intersection of the two sets.

Given a periodic array of N-elements, the array factor $F(u)$ is:

$$F(u) = \sum_n c_n e^{jkndu} \qquad (3.17)$$

where it is understood that certain external constraints might be imposed on the coefficients c_n (such as a limitation to the dynamic range or the phase progression between adjacent elements, etc.).

The set B contains all possible functions $F(u)$, and when constraints are put on the excitation they define the subset B_c .

The requirements on the pattern are put in the form of two masks, an upper bound $M_U(u)$ and a lower bound $M_L(u)$ such that the required patterns must fall on or between these two bounds, and the set of all such patterns is the set M. An array factor that belongs to both sets M and B_c is a solution to the synthesis problem. Figure 3.5(a) shows the mask set for a particular case of a flat-topped radiation pattern with sidelobe below the -20 dB level, and decaying at larger angles. The synthesis is complete when a function within set B_c is also within the mask set M.

The term alternating projection refers to the use of the concept of successive projectors. A projector is an operation that gives the best possible approximation to some function subject to a chosen norm, like the mean square norm (L^2). In this case the Fourier series is known to give the best mean square approximation to a given function, and is the basis for the alternating projection scheme.

The iterative process leads from an approximation of the pattern function x_n to the next iterated pattern function x_{n+1} by means of two projectors using the iteration

$$x_{n+1} = P_B P_M x_n \qquad (3.18)$$

where P_B and P_M are projection operators applied in the sequence shown. The projector operation P_M described by the mask set simply restricts the pattern to be between the upper and lower masks. Starting with a radiation pattern $F(u)$, the sequence begins below:

$$P_M F(u) = \begin{cases} M_U(u)\frac{F(u)}{|F(u)|} & |F(u)| > M_U(u) \\ F(u) & M_L(u) \le F(u) \le M_U(u) \\ M_L(u)\frac{F(u)}{|F(u)|} & |F(u)| \le M_L(u) \end{cases} \qquad (3.19)$$

Having forced the pattern to be within the bounds of the mask, the next series of actions is the projection P_B with constraints on the currents, and is shown below.

$$P_{B_c} = \bar{f}_s f_c f_N \bar{f}_s^{-1} \qquad (3.20)$$

This projector over the excitation constrained set B_c alters the currents in the following way. Taken from right to left, the inverse transform operating on this new field pattern arrives at a new set of currents. These are operated upon by a process denoted by f_N that sets to zero all Fourier coefficients that are outside of the array. Next, if desired, is the application of f_c, a constraint operator that may be imposed on the remaining Fourier series terms (the currents) to limit their dynamic range or for other purposes. Then using the new constrained currents the Fourier series operator $\bar{f}_s$ evaluates the far field pattern corresponding to this iteration.

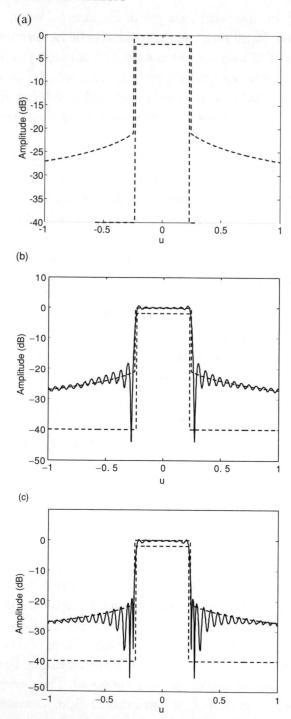

FIGURE 3.5: Alternating projection synthesis using 64 elements, (a) The upper and lower masks and the initial estimate, (b) Result of first iteration, and (c) Result of tenth iteration.

With these two projectors P_M and P_{B_c}, the iterative sequence is defined by Eqs. (3.18) and (3.19). Amplitude constraints are imposed with f_c by merely forcing the Fourier series coefficients to be constrained. For example, to require the dynamic range to be constrained to be within $c_{min} \leq |c_n| \leq c_{max}$, the f_c maps a sequence a_n into b_n using:

$$b_n = \begin{cases} c_{max} \dfrac{a_n}{|a_n|} & |a_n| > c_{max} \\ \\ = \{ a_n & c_{min} \leq a_n \leq c_{max} \\ \\ = \left\{ c_{min} \dfrac{a_n}{|a_n|} \right. & |a_n| < c_{min} \end{cases} \qquad (3.21)$$

Similarly one can impose a maximum phase variation between elements or other constraints.

Convergence with this system is not guaranteed, for M and possibly B_c are not convex sets, so it is important that the initial guess at the pattern be reasonably good, otherwise the sequence could converge to a local minimum.

Figure 3.5 shows several iterations of a sequence that demonstrates the application of this technique for generating a flat pulse radiation pattern with decaying sidelobes using 64 elements with half wave spacing. Throughout these curves the upper and lower masks are shown dashed. Instead of approximating the pattern, an initial pattern was chosen as unity over all space, and its projection P_M onto the masks using Eq. (3.19) made it coincide with the upper mask boundary (Fig. 3.5a). Figure 3.5(b) shows this far field for the first iteration, and Fig. 3.5(c) shows that the result of the tenth iteration is a pattern narrowed to fit within the masks with only slight deviations outside of the mask.

With more complicated patterns it is often necessary to take more care with the initial estimate. Lack of convergence usually means that the chosen mask did not contain a physically realizable pattern.

REFERENCES

[1] H. Steyskal and J Herd, " Mutual coupling compensation in small array antennas", *IEEE Trans.*, Vol. AP-38, No. 12, pp. 1971–1975, Dec. 1990.

[2] L. Pettersson, M. Danestig and U. Sjostrom, An experimental S-band digital beamforming array," *1996 IEEE International Symposium on Phased Array Systems and Technology*, 15–18 Oct. 1996, Boston MA.

[3] R. J. Mailloux, "Operating Modes and Dynamic Range of Active Space Fed Arrays with Digital Beamforming," Vol. AP-54, No. 11, pp. 3342–3355, Nov. 2006.

[4]   P. M. Woodward, "A method of calculating the field over a plane aperture required to produce a given polar diagram," *Proc. IEE, Part 3A*, Vol. 93, pp. 1554–1588, 1947.

[5]   P. M. Woodward and J. P. Lawson, "The theoretical precision with which an arbitrary radiation pattern may be obtained from a source of finite size," *Proc. IEE*, Vol. 95, P1, pp. 362–370, Sept. 1948.

[6]   C. L. Dolph, "A current distribution for broadside arrays which optimizes the relationship between beamwidth and sidelobe level," *Proc. IRE*, Vol. 34, pp. 335–345, June 1946.doi:10.1109/JRPROC.1946.225956

[7]   R. J. Stegen, "Excitation coefficients and beamwidths of Tschebyscheff arrays," *Proc. IRE*, Vol. 41, pp. 1671–1674, November 1953.doi:10.1109/JRPROC.1953.274197

[8]   T. T. Taylor, "Design of line source antennas for narrow beamwidth and low sidelobes," *IEEE Trans. Antennas Propag.*, Vol. AP-3, pp. 16–28, Jan. 1955.

[9]   T. T. Taylor, "Design of circular apertures for narrow beamwidth and low sidelobes," *IRE Trans. Antennas Propagation*, Vol. 8, pp. 17–22, Jan. 1960.doi:10.1109/TAP.1960.1144807

[10] R. C. Hansen, "Linear Array," in A.W. Rudge et al., Eds., *The Handbook of Antenna Design*, Vol. 2, Peter Peregrinus, London, 1983, Chap. 9, p. 30.

[11] C. F. Winter "Using Continuous Aperture Illuminations Discretely," *IEEE Trans.*, Vol. AP-25, Sept. pp. 695–700, 1977.

[12] R. S. Elliott, On discretizing continuous aperture distributions," *IEEE Trans.*, Vol. AP-25, Sept. 1977, pp. 617–621.

[13] E. T. Bayliss, "Design of monopulse antenna difference patterns with low sidelobes," *Bell Syst. Tech. J.*, Vol. 47, pp. 632–640, 1968.

[14] Y. Rahmat-Samii and E. Michielssen, *Electromagnetic Optimization by Genetic Algorithms*, chapter 5. Wiley, 1999.

[15] D. Marcano and F. Duran, "Synthesis of antenna arrays using genetic algorithms," *IEEE, AP-S Magazine*, Vol. 42, No. 3, June 2000, pp. 12–18.

[16] R. J. Mailloux, "A low sidelobe partially overlapped constrained feed network for time-delayed subarrays", *IEEE Trans.*, AP-49, No.2, Feb. 2001, pp. 280–291.

[17] O. M. Bucci et al., "Intersection approach to array pattern synthesis", *IEE Proc.*, Vol. 137, Pt. H, pp. 349–357, Dec. 1990.

CHAPTER 4

Subarray Techniques for Limited Field of View and Wide Band Applications

4.1 INTRODUCTION

The thinned array concept discussed in the section on synthesis provides a simple way to form a very narrow scanning beam with a reduced number of elements. A thinned array has ensemble average sidelobes at the $1/N$ level, where N is the number of active elements in the array. Peak sidelobes will be many dB above the average levels, but this may be acceptable depending upon specifications. Thinned arrays have reduced gain however, and so may not be applicable to systems that require high aperture efficiency or where space is limited. In these cases it is appropriate to use fully filled array apertures, usually with a periodic lattice.

Several array applications require subarray level architecture to reduce the number of active controls, devices, and signal processing. These applications are for arrays that scan over a limited field of view or for relatively wide band arrays that use time delay control behind each subarray, and phase shift in the aperture. These two applications are depicted in Fig. 4.1. Since they require the same subarray technology they are treated together here for illustrative purposes.

Figure 4.1 shows an array of microstrip patch elements and two kinds of subarrays suited to these applications. Limited field of view arrays (Fig. 4.1(a)) include satellite communication systems, airport approach radars, weapon guiding systems in addition to other applications. These systems are required to scan only a few degrees (perhaps to 5 or 10°), and so can use phase shifters behind each subarray instead of behind each element. Figure 4.1(b) applies to large arrays with bandwidths that are beyond the limits of the squint bandwidth (Eq. 1.16) for the given size array. Such systems include large ground and space based radar arrays and satellite communication systems. For these systems, Fig. 4.1(b) indicates that time delay is applied at the input terminals and phase shift at the elements.

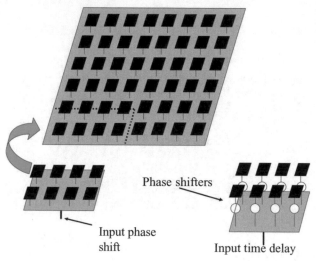

Phase shifters

Input phase
shift

Input time delay

FIGURE 4.1: Array showing subarrays for limited field of view (LFOV) and wide band operation: (a) subarray for LFOV and (b) subarray for time delay.

One obvious solution to both requirements is to use equal signal power dividers as subarrays. Unfortunately this approach produces contiguous subarrays that have significant "quantization" sidelobes. An example is shown in Fig. 4.2 for a linear array of 512 elements using time delayed subarrays of 32 elements each. In this case even a small frequency offset of 2.5% produces very large quantization lobes. The 'quantization lobes' occur at the grating lobe distances $u_p = u_0 + p\lambda/D_x$ and $v_q = v_0 + q\lambda/D_y$ where D_x and D_y are the distances between subarrays. They are called quantization lobes instead of grating lobes since they are caused by quantizing the amplitude, phase or time delay at each subarray and unlike the grating lobes they are smaller than the main beam.

Figure 4.3 illustrates how the quantization lobes are formed in the limited field of view (LFOV) and time delay cases. Figure 4.3a and 4.3c show a generic phased array of length L that requires discrete phase shift from 0 to $2\pi(L/\lambda \sin\theta)$ for time delay from 0 to $\frac{L\sin\theta}{c}$ for scan to the angle θ. The basic array of Figures 4.3b and 4.3d has 64 elements, separated into eight element subarrays of half wave spaced elements, and the array is uniformly illuminated. Figure 4.3(a) illustrates the cause of these quantization lobes for an LFOV array. The array requires a linear progressive phase shift, but since there are no phase shifters in the subarrays, the actual element phase is a staircase of discrete phase steps, spaced D_x apart. The resulting sketch of phase errors is shown at the right of the figure. Since this spacing is greater than a wavelength, there results a series of quantization lobes in the array factor as shown in the middle two plots of Fig. 4.3(b). The left three sketches in Fig. 4.3(b) are the broadside behavior of the subarray (top), array factor (center), and array pattern (bottom), these are also shown in the right three patterns for the array scanned to $u_o = 0.1$ (5.74°). The subarray patterns do not

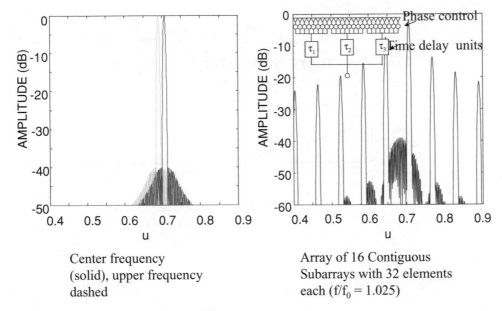

Center frequency
(solid), upper frequency
dashed

Array of 16 Contiguous
Subarrays with 32 elements
each ($f/f_0 = 1.025$)

FIGURE 4.2: "Squint" property of large phased arrays.

change with scan but the array factors move so that the main beam peak is at u_0. On the top left is the eight element subarray pattern with maximum at broadside and nulls at $u = \pm 0.25$. The central pattern, the array factor for broadside scan, has grating lobes at $u_p = p\lambda/D_x$, or at the direction cosines $-.25$, 0, and $.25$, with the main beam at 0. The lower left pattern is the array radiation pattern, the product of the subarray pattern and the array factor. Note that the array element pattern has been omitted to emphasize the effect of the subarray pattern. The product pattern at the lower left is the ideal p-sinc pattern of the uniformly illuminated array with the grating lobes totally suppressed by the subarray pattern nulls. The three figures at right show the array scan behavior with the subarray pattern unchanged (there are no phase shifters in the subarray) and the array factor scanned with peak at $u_o = 0.1$. Now the array radiation pattern at bottom shows significant quantization lobes as well as a reduction of the peak value. The lobes are present since the subarray pattern nulls are stationary throughout scan and no longer cancel the quantization lobes.

The situation is very similar for the array using time delay at the subarrays to produce a "wideband" scanned beam over a wide angle of scan. In this case the same array geometry uses phase shifters at every element, but time delay at the subarray input ports (Fig. 4.1b). Figure 4.3(c) shows a wave incident on the array at angle θ and the straight line is the required time delay. Below that sketch is the applied phase shift imposed at every element to match the slope of the required time delay curve. The lower right sketch shows that when time delay is added, the result is a saw tooth approximation to the correct time delay. The residual error leads to the array pattern lobes of Fig. 4.3(d).

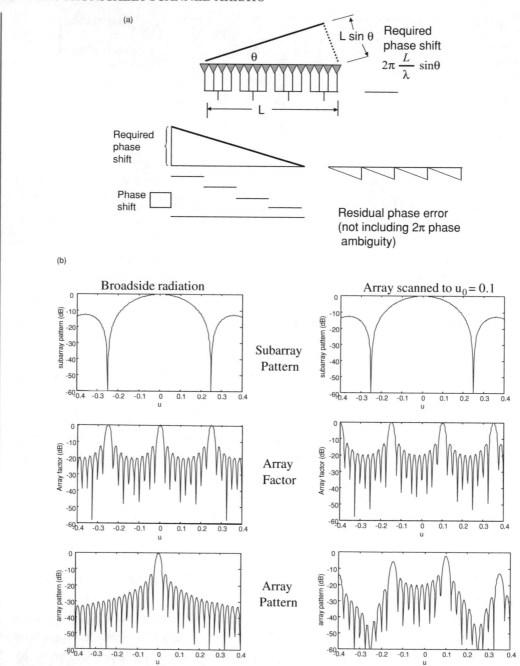

FIGURE 4.3: Subarray and array patterns showing quantization lobe formation for LFOV and wide-band arrays: (a) Broadside subarrays for LFOV scanning, (b) subarray patterns, array factors and radiation patterns for LFOV array, (c) phase shifted subarrays for wideband time-delay steered array, and (d) Subarray patterns, array factors and radiation patterns for time-delay steered array.

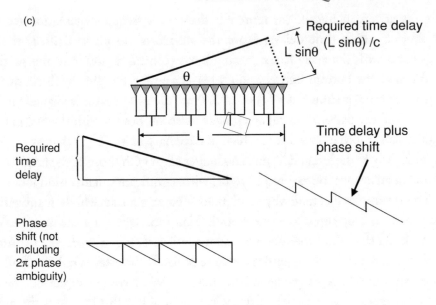

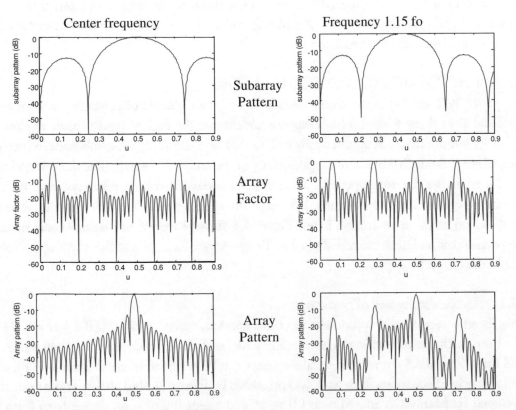

FIGURE 4.3: Figure 4.3 (continued).

The plots at left in Fig. 4.3(d) represent the center frequency situation for the array scanned to $u_0 = .5$ using time delays. Since the subarrays are phase shifted, the subarray pattern nulls align with the array factor grating lobes (central figure), and the product array radiation pattern at the bottom left, is again a perfect p-sinc function with no quantization lobes. At right the array is shown at f/f_0 at 1.15. Now the array factor is virtually unchanged, its peak is at u_0, and the pattern has imperceptively narrower beams, with the two grating lobes moving slightly closer to the main beam which is stationary at u_0. The subarray patterns have squinted significantly to the left and its nulls no longer align with those of the array factor. This product results in an array pattern (bottom right) with significant quantization lobes.

These two examples illustrate why contiguous subarrays produce high quantization lobes. The subarray pattern only suppresses the quantization lobe when the subarray peak is a the array beam peak. At that point the subarray pattern nulls are at $u = u_0 \pm p\lambda/D_x$, and exactly coincide with the locations of the quantization lobe peaks. If the array is uniformly illuminated there is no remaining vestige of the quantization lobes. We have not discussed the situation wherein the array taper is done at the subarray input ports, but this broadens the array factor beams and then even if the nulls do coincide with those of the subarray patterns, there will remain vestigial quantization lobes. Amplitude quantization produces a second-order effect compared to the results discussed above.

4.2   BASIC SUBARRAY APPROACHES

The paper by Tang [1] is the classic description of the methods of producing subarrays for wideband time delay applications. Tang did not discuss the limited field of view application explicitly, but the methods apply equally well to that case. Basically these methods fit into two categories; methods that break up the periodicity of the array of subarrays (like the thinned array mentioned earlier) by using aperiodic subarrays or interlaced irregular subarrays, and methods that use a periodic array of subarrays, but shape the subarray patterns to suppress radiation in the direction of the quantization lobes. Figure 4.4 shows some of the various basic subarray approaches that had been conceived prior to Tang's paper. Examples of these will be presented later in the chapter.

4.2.1   Aperiodic Arrays of Subarrays

Tang listed several classic approaches to utilize aperiodic arrays of identical subarrays. These have been further exploited in the intervening years, and some of the basic early references are cited here. Figure 4.5 shows that equal subarrays can be arranged in circular rings to form an aperiodic grid of subarrays. This work was presented by Patton [2] and almost at the same time developed by Manwaren and Minute [3] who used unequal size large elements to form the aperiodic planar array.

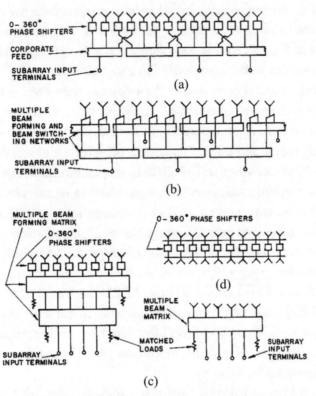

FIGURE 4.4: Subarray schemes for minimizing gain reduction and grating (quantization) lobe level (after Tang [1]). (a) Interlaced subarray scheme,(b) overlapped subarray scheme,(c) completely overlapped subarray scheme (constrained feed approach), and(d) completely overlapped subarray scheme (space feed approach).

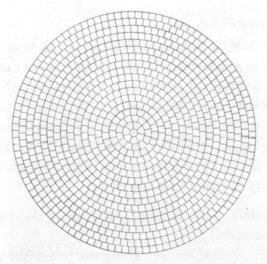

FIGURE 4.5: Circular planar array of subarrays (after Patton [2]).

Recently this author and colleagues [4] have investigated the use of one or several types of irregular polyomino shaped subarrays. A polyomino is a figure formed by interconnected squares as indicated in Fig. 4.6. In this study the figures are rotated by 90 intervals, or flipped to produce an array of subarrays that completely fills a periodic rectangular grid aperture. In order to feed the array with practical components the study has been limited to subarrays with four (tetromino) or eight (octomino) elements, and only one type of subarray (or one type plus the flipped version) is used in an array. Initial results have shown that in comparison to that for an array of contiguously spaced subarrays, this scheme achieves additional peak quantization lobe suppression of $10-19$ dB for arrays of 128–2048 L-shaped octomino subarrays. In this case the larger array, when compared to an array with eight element rectangular subarrays, eliminated the -12 dB quantization lobe, and reduced all remaining lobes to below -31 dB. Moreover, unlike the quantization lobes or periodic arrays, the residual peak sidelobes of the new arrays continue to decrease as the number of subarrays is increased. The technique can therefore have much lower sidelobes for larger arrays, but as yet there are only theoretical results to confirm this performance. Figure 4.6(b) shows an array of 256 L-octomino shaped subarrays (2048 elements). Figure 4.6(c) compares the quantization lobes of the array of rectangular 8-element subarrays with the peak sidelobes of the array filled with L-octomino shaped subarrays. The new array had maximum sidelobes of -25 dB, more than 14 dB below the -11.5 dB quantization lobe of the array of rectangular subarrays.

Other recent studies of irregular tiled arrays include the work of Pierro et al. [5] who obtained general relationships about operating bandwidth and sidelobe spectrum for a variety of array tilings, and the study of Kirby and Bernhard [6] who looked at making very large arrays from large identical random subarrays and rotating adjacent ones to reduce the periodicity and lower the sidelobe level of the array factor.

4.2.2   Interlaced Subarray Technology

Figures 4.4(a) and 4.7 illustrate the technique of interconnecting elements to form an interlaced subarray. Figure 4.7 shows a square grid array lattice with different shading to indicate that the elements share common subarrays. This technique as presented by Stangel [7] uses subarray spacing that form a rectangular lattice of subarray centers, but with each subarray an irregular shape interlaced with other local subarrays. Figure 4.7(b) shows the reduction in phase shifter elements achieved for scan angles up to 40° with maximum sidelobe level -15 dB. Clearly this technique is capable of achieving lower sidelobes for a larger array, since the irregular shape and spacing of the subarrays approximates a random array.

4.2.3   Pulse Shaped Subarray Pattern for Quantization Lobe Suppression (Overlapped subarrays)

Before discussing specific configurations, it is illustrative to consider a special subarray pattern, used in a periodic array that would eliminate all quantization lobes. For subarrays of length

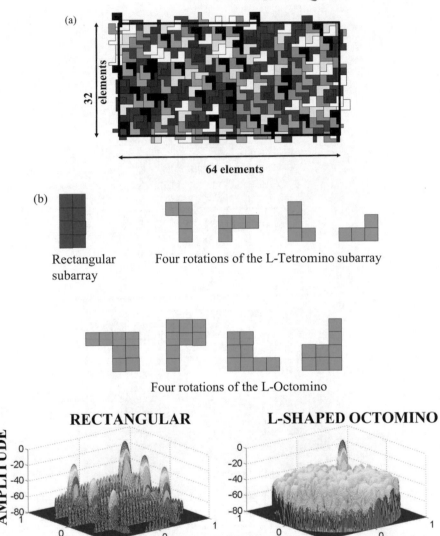

(a)

32 elements

64 elements

(b)

Rectangular subarray

Four rotations of the L-Tetromino subarray

Four rotations of the L-Octomino

(c)

RECTANGULAR

L-SHAPED OCTOMINO

AMPLITUDE

V U

AMPLITUDE

V U

FIGURE 4.6: Irregular (polyomino shaped) subarrays: (a) L-tetromino and L-octomino shaped subarrays, (b) array of L-shaped octomino subarrays, and (c) comparison of sidelobe levels of rectangular and L-shaped subarrays (after Mailloux [4]).

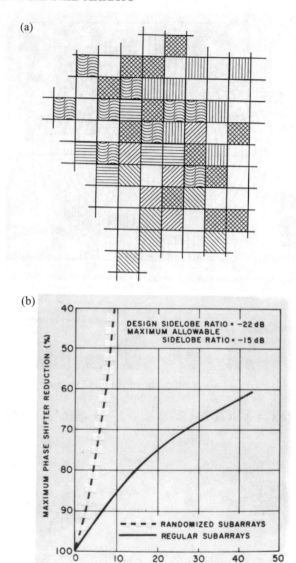

FIGURE 4.7: Interlaced subarrays. (a) Interlaced geometry, (b) phase shifter reduction (after Stangel et al. [7]).

D_x in one plane, and for either the time delay or LFOV case, if a very narrow main beam is formed at some angle u_0 (zero for the LFOV case), then the nearest quantization lobes are at the direction cosines $u_0 + \lambda/D_x$ and $u_0 - \lambda/D_x$. A pulse shaped subarray pattern that had unity amplitude for $u_0 - 0.5\lambda/D_x \leq u \leq u_0 + 0.5\lambda/D_x$ would therefore allow for maximum scan with complete quantization lobe suppression.

For a very large array with beamwidth much less than the subarray pattern width, Fig. 4.8 shows a flat-topped pulse-type subarray pattern that would have the ideal shape to suppress the quantization lobes. Then, whether the array factor pattern is scanned by the input signals as in the LFOV case, or whether the subarray pattern squints in the wide band mode, all quantization lobes are located within the subarray pulse shaped stop band.

Minimum Number of Control Elements

The flat-topped subarray pattern of Fig. 4.8 helps to explain that there is a minimum number of controls for both the time delay and limited field of view cases.

For the LFOV case Fig. 4.8 shows that the array can scan to $\sin \theta = 0.5\lambda / D_x$ without quantization lobes. For a linear array of length L with beamwidth θ_3, organized into subarrays of length D_x, and within a flat-topped region $\theta \leq |\theta_{max}|$, the approximate minimum number of subarrays N_{min} is readily shown to be:

$$N_{min} = \frac{L}{D_x} = \frac{2 \sin \theta_{max}}{\sin \theta_3}. \tag{4.1}$$

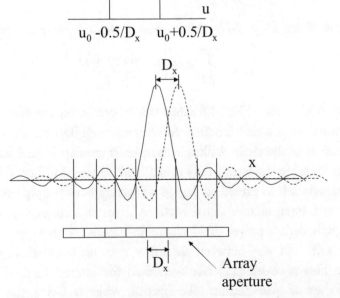

FIGURE 4.8: Ideal subarray pattern (top) and overlapped sinc function aperture distributions (bottom).

This expression has a useful interpretation as the number of beams to fill the space $-|\theta_{max}| \leq \theta \leq |\theta_{max}|$. One can extend this logic to the two-dimensional case and to scanning lenses and reflectors, and a number of excellent works have fully generalized the relationship [2, 8, 9].

Applying the same constraint applied to the fractional bandwidth limitations for a phased array scanned to the direction cosine u_0 with time-delayed subarrays leads to:

$$\frac{\Delta f}{f_0} = \frac{\Delta u}{u_0} = \frac{\lambda}{D_x \sin \theta_0}. \tag{4.2}$$

These ideal bounds cannot be reached in practice, even for a very narrow main beam because any real subarray pattern cannot duplicate the vertical pulse sides, but one can often approach 60–80% of this value.

There have been many constrained networks developed to produce the overlap required for both of these applications. Most of these are described in the comprehensive paper by Skobelev [10], but a limited few are described below.

4.2.4   Overlapped Subarray Synthesis of Flat-Topped Patterns

The array illumination required to produce (or approximate) the pulse shaped subarray pattern at the top of Fig. 4.8 is obtained from the Fourier synthesis technique as:

$$a_n = \frac{d_x}{\lambda} \int_{u_0-(\lambda/2D_x)}^{u_0+(\lambda/2D_x)} F(u)e^{-j\frac{2\pi}{\lambda}nud_x}du \tag{4.3}$$

For a subarray of size $D_x = Md_x$, this reduces to the following sinc function:

$$a_n = \frac{1}{M}e^{-j2\pi n\frac{d_x}{\lambda}u_0}\frac{\sin\left(\pi\frac{n}{M}\right)}{\left(\pi\frac{n}{M}\right)} \tag{4.4}$$

The sketch at the bottom of Fig. 4.8 shows that to synthesize the flat-topped distribution the necessary illumination is a sinc function and extends well beyond one subarray, and even a rough approximation to the desired illumination must overlap several subarrays. This fact explains the reason for using the term " fully overlapped" as noted in Fig. 4.4.

The simplest network to provide a degree of overlap is to simply sum the signals from adjacent elements and interpolate a signal between them as indicated in Fig. 4.9. The figure shows two signals with a power divider/combiner network that produces an additional signal with phase shift half way between the other two but (normalized) amplitude modulated by $\cos \delta/2$. This network [11] has been used for limited field of view applications for providing a degree of quantization lobe control, while reducing the number of phase shifters by a factor of two. Far more sophisticated networks have been devised to provide the

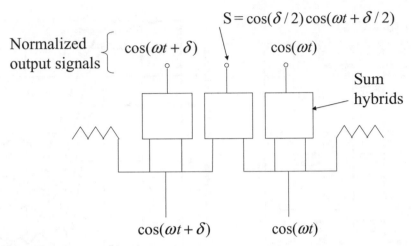

FIGURE 4.9: Simple overlapped subarray using power dividers (after Mailloux [11]).

overlap necessary to approximate a flat-topped subarray pattern. Among these research studies are the works or Skobelev and colleagues [12, 13], and others [14,15]. Figure 4.10 shows an analog nework invented by Skobelev [13], for providing overlap, and the subarray patterns for varying cascades of the Skobelev Chess network. These kinds of constrained overlap network are partially overlapped systems, because they do not span the whole array, and only connect to a few neighbor elements. In practice they tend to suppress quantization lobes to about the −20 dB level. Recent work by Herd et al. [15] included making a printed circuit overlap network.

Dual Transform Networks for Overlapped Subarrays

In the early 1960s the Hughes Corporation developed a unique wideband scanning system shown schematically in Fig. 4.11, called HIPSAF [16] (High Performance Scanned Array Feed) to synthesize completely overlapped subarrays. The system is based on the principle of cascading two multiple beam networks that serve as a telescope. One network was a focusing lens and the other a Butler matrix. The figure also illustrates the principle of operation for this system. As each port of the Butler matrix is excited, the network forms a constant amplitude, progressive phase at its radiating output ports. This radiation has a p-sinc pattern that illuminates the back face of the lens and is collimated and phase shifted, and fed through to the radiating front face of the lens where it radiates a beam with the required pulse shape. Each of the Butler matrix input ports controls a pulse shaped subarray with time delay at its input. The output of each network is an approximate Fourier transform of the input ports, and so this type of system is often called a dual transform beamformer.

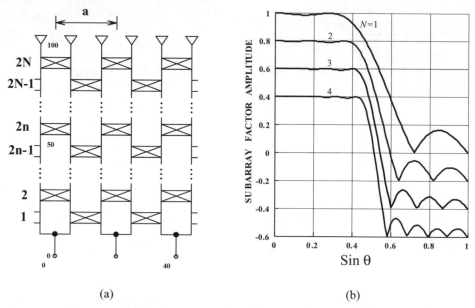

(a)   (b)

FIGURE 4.10: Chess network of Skobelev (after Skobelev [13]).

Figure 4.12 shows two such transform networks. At left are two cascaded Butler matrices that form a dual transform, while at right is a network like HIPSAF consisting of a focusing lens and a multiple beam feed. Similar configurations using a multiple beam feed and a reflector objective can also be used but will not be described here. It should be mentioned here that the smaller $M \times M$ matrix might (and in many cases probably will) be done digitally. For the purposes of analysis we will consider a dual transform network of two back-to-back orthogonal (Butler) matrices.

The dual transform type of overlapped subarray system is best understood as a transmit system. Figure 4.12 depicts the two orthogonal multiple beam systems in cascade. If the $M \times N$ Butler matrix ($N > M$) is excited by a signal I_i applied at the ith port, then the output signals A_n (which are the aperture currents of the radiating array) are [17]

$$A_n = I_i \frac{1}{\sqrt{N}} \exp^{[-j2\pi i(n/M)]} \tag{4.5}$$

which radiate as p-sinc functions:

$$g_i(u) = I_i N^{1/2} f^e(u) \frac{\sin[(N\pi d_L/\lambda)(u - u_i)]}{N\sin[(\pi d_L/\lambda)(u - u_i)]} \tag{4.6}$$

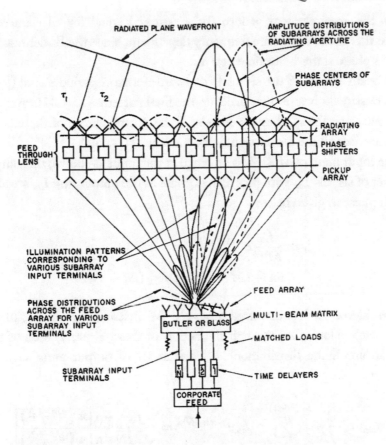

FIGURE 4.11: Completely overlapped subarray antenna system (HIPSAF) (after Tang [1]).

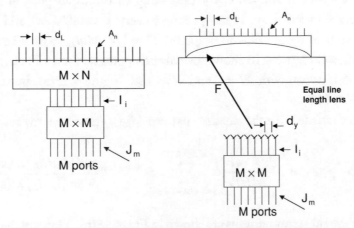

FIGURE 4.12: Constrained and space fed overlapped subarray beamformers.

where $f^e(u)$ is the array element pattern (here assumed equal for all elements) and $u_i = i\lambda/(Nd_L)$. Note that the u_i angles are frequency dependent, since the Butler matrices radiate a wavefront that is phase shifted, not time delayed.

This ith beam is one of the set of Woodward–Lawson periodic sinc (p-sinc) beams. These are the 'constituent beams' that make up the final pattern. Only M beams are accessible here, so that means that the synthesis is done with the central cluster of these beams (with an $M \times M$ set of matrices all the beams would of course be accessible).

When the input matrix $(M \times M)$ is attached at the input as shown, then any single input J_m produces a set of signals I_{im} with progressive phase at the output, the I_{im} are all of the same magnitude with phase as given below

$$I_{im} = \frac{J_m}{M^{1/2}} e^{-j2\pi(m/M)i} \tag{4.7}$$

$$\text{for} - (M-1)/2 \leq i \leq (M-1)/2.$$

This input signal J_m has therefore produced M different signals I_{im}, one at each ith output port for each value of m, through the action of these 'i' ports, each of these J_m signals produces an amplitude distribution A_{mn} at the set of output ports such that for any m,

$$A_{mn} = \frac{1}{\sqrt{N}} \sum_{i=-(M-1)/2}^{i=(M-1)/2e} I_{im} e^{j2\pi(n/N)i} = \frac{J_m}{\sqrt{MN}} \frac{\sin\left[\pi\left(\frac{nM-mN}{N}\right)\right]}{\sin\left[\pi\left(\frac{nM-mN}{NM}\right)\right]} \tag{4.8}$$

This is the aperture illumination corresponding to the input port, and is the aperture illumination of the mth subarray. It has peaks at every $n = m(N/M)$ and so the aperture illumination is a sum of p-sinc functions spaced $D = (N/M)d_L$ apart where D is thus the inter-subarray distance. Figure 4.13 shows several such patterns for a 128 element array fed by a Butler matrix combination with $M = 8$ and $N = 128$. Figure 4.13(a) shows the M subarray amplitudes from Eq. (4.8).

These signals radiate with the subarray patterns $f_m(u)$ that have the form

$$f_m(u) = \frac{1}{J_m} \sum_i g_i(u) = N \frac{f^e(u)}{(MN)^{1/2}} \sum_{i=-(M-1)/2}^{i=(M-1)/2} e^{j2\pi(m/n)i} \frac{\sin[(N\pi d_L/\lambda)(u - u_i)]}{N\sin[(\pi d_L/\lambda)(u - u_i)]} \tag{4.9}$$

Several of these subarray patterns are shown in Fig. 4.13(b). They are the sum of M of the constituent (Woodward) beams from Eq. (4.6). These are the subarray patterns corresponding to the input J_m, for $m = 1$ and 3 and have the (rippled) flat top shape with the approximate width given below

$$\Delta = \frac{M}{N}\left(\frac{\lambda}{d_L}\right). \tag{4.10}$$

The reason for this subarray width is that the pattern combines M constituent beams, each one of width $1/(Nd_L)$. Notice that the pattern for the subarray ($m = 1$) nearer the array edge has significant ripples.

When all of the input M ports are excited with a tapered distribution as appropriate for a low sidelobe pattern, the combined amplitude distribution at the radiating array is shown

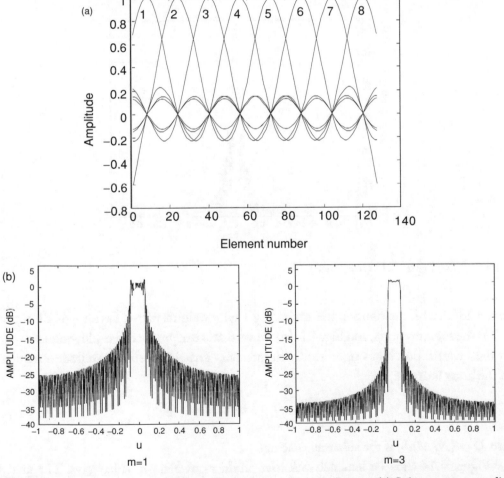

FIGURE 4.13: Constrained completely overlapped subarray beamformer. (a) Subarray aperture distribution, (b) subarray pattern, (c) array aperture distribution, and (d) radiation pattern.

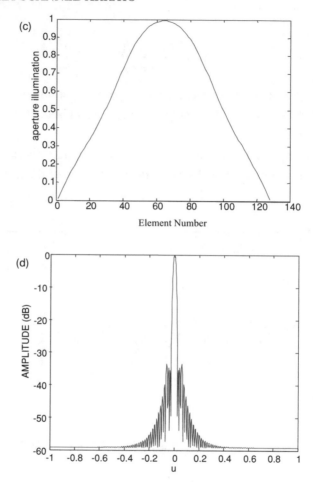

FIGURE 4.13: Figure 4.13 (continued).

in Fig. 4.13(c). This distribution was excited by Taylor weights with a Taylor -40 dB taper and $\bar{n} = 7$ at the subarray ports, and Fig. 4.13(d) shows that it can produce a very low sidelobe pattern with the quantization lobes significantly suppressed. The time delayed excitation required to scan the array beam is:

$$J m = |J_m|\, e^{-j2\pi n(D/\lambda_0)u_0} \qquad (4.11)$$

where $D = (N/M)d_L$ is the subarray spacing.

Figure 4.14 shows a lens network that produces overlapped subarrrays. The analytical description of this network is given in [17], but some interesting general factors are noted on Fig. 4.14. The angleΔ, subtended by the multiple beam feed is also the angular width of the

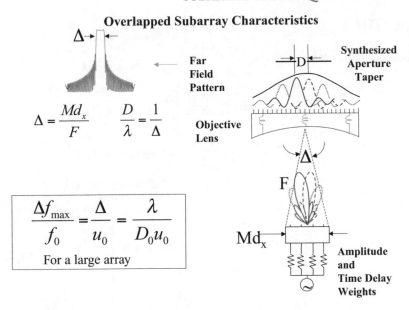

Overlapped Subarray Characteristics

$$\Delta = \frac{Md_x}{F} \qquad \frac{D}{\lambda} = \frac{1}{\Delta}$$

$$\frac{\Delta f_{max}}{f_0} = \frac{\Delta}{u_0} = \frac{\lambda}{D_0 u_0}$$

For a large array

FIGURE 4.14: Overlapped subarray characteristics (illustrated for space fed lens).

subarray pattern (shown upper left), and is also the inverse of the normalized subarray spacing D/λ. This makes the approximate fractional bandwidth $\Delta f_{max}/f_0$ equal to the ratio of $\Delta u/u_0$ where Δ is the subarray pattern beamwidth. Note that this bandwidth is M times that of a single beam radiating from the face to the lens. In addition to showing the subarray pattern, Fig. 4.14 also shows the p-sinc functions combining to form the aperture distribution (it performs an approximate Woodward synthesis). Early studies of this configuration are the work of Borgiotti [18], Fante [19], and a detailed experimental study by Southall and McGrath [20].

Now assume the whole set of signals J_m with progressive phase are applied at the input of the $M \times M$ matrix. If that phase matches with one of the orthogonal beams, and the J_m are all the same amplitude, then all the signal combines into one port I at the $M \times M$ output, and it radiates a single constituent beam. If the J_m do not exactly correspond to an orthogonal progression, then a group of constituent beams are formed. If the amplitude of the set J_m is tapered, then the proper combination of constituent beams are combined to form a low sidelobe radiated array factor.

The techniques described for forming the fully overlapped subarrays using Butler matrices and lenses (or reflectors) are intended for very large arrays with relatively narrow band applications compared to the network overlap schemes [6–10] that overlap only a few elements. Here the technique works well with subarrays separated by many wavelengths. These techniques can in principle lead to very low sidelobes whose height is determined primarily by component tolerance, array coupling, and edge effects. A primary advantage of the fully overlapped technique

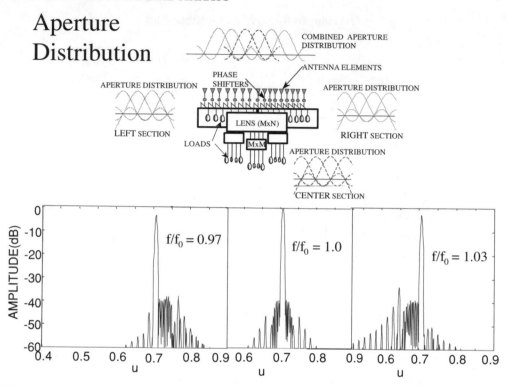

FIGURE 4.15: Partially overlapped subarrays (after Mailloux [21]).

is the very large reduction in the number of controls compared to element level control of these large structures.

Figure 4.15 shows a scheme [21] that addresses the issue of how to provide high quality overlapped subarrays for a very large linear array. The technique uses moderate size dual transform systems (with one transform perhaps being done digitally) to construct a modular array by incorporating additional lenses as noted in Fig. 4.15. The basic issue is that to build a large lens that scans in one dimension, one needs to constrain the power division with a parallel plate structure, and for a fixed bandwidth the ratio of depth F to aperture length L is constant. So if the length L increases, so must F, until the structure becomes too deep and unwieldy. The sketch of Fig. 4.15 shows three sections of an array of many more sections. Each section is a dual transform network (probably a digital transform and a Rotman lens). If each section forms (for example) four subarrays as shown, these would have aperture illuminations that would overlap within each section. Of the four, two of each network are well formed (free of edge effects), and so are used to provide quality subarray patterns. The unwanted subarrays in each section are terminated in matched loads and the higher quality subarrays are added together by partially overlapping the adjacent section outputs. Additional sections can be added, making this a modular construction that has very shallow depth, and suitable for a long one-dimensional array or line source feed for a two-dimensional structure. Results indicate that sidelobe levels in the area of −30 dB can be obtained with this simple scheme.

REFERENCES

[1]   R. Tang, "Survey of time delayed beam steering techniques," in *Phased Array Antennas: Proc. of the 1970 Phased Array Antenna Symposium*, Artech House, Dedham, MA, pp. 254–260, 1972.

[2]   W. Patton, "Limited scan arrays," *in Phased array antennas: Proc . 1970 Phased Array Symposium*, A.A. Oliner and G.A. Knittel, Eds. Artech House 1972, pp. 254–270.

[3]   T. A. Manwarren and A. R. Minuti, "Zoom feed technique study,", RADC-TR-74–56, Final technical report 1974.

[4]   R. J. Mailloux, S. G. Santarelli and T. M. Roberts, "Wideband arrays using irregular (polyomino) shaped subarrays", *Electr. Lett.*, Vol. 42, No 18, pp. 1019–1020, 31 Aug. 2006.doi:10.1049/el:20062252

[5]   V. Pierro, G. Galdi, G. Castaldi, I. M. Pinto and L. B. Felson, "Radiation properties of planar antenna arrays based on certain categories of aperiodic tilings", *IEEE Trans.*, Vol. AP-53, No. 2, pp. 635–664, Feb. 2005.

[6]   K. C. Kirby and J. T. Bernhard, "Sidelobe level and wideband behavior of arrays of random subarrays," *IEEE Trans.*, AP-54, No. 8, pp. 2253–2262, Aug. 2006.

[7]   J. Stangel and J. Ponturieri, "Random subarray techniques," *IEEE G–AP Int. Symp.* Dec. 1972.

[8]   G. Borgiotti, "Degrees of freedom of an antenna scanned in a limited sector," *IEEE International Symp.*, pp. 319–320, 1975.

[9]   J. Stangel, "A basic theorem concerning the Electronic scanning capabilities of antennas," URSI Commission 6, Spring meeting, June 11, 1974.

[10] S. P. Skobelev, "Methods of constructing optimum phased-array antennas for limited field of view", *IEEE Antennas Propagation Mag.*, Vol. 40, No. 2, pp. 39–49, April 1998.doi:10.1109/74.683541

[11] R. J. Mailloux , P. R. Caron, "A class of phase interpolation circuits for scanning phased arrays," *IEEE Trans.*, Vol. AP-18, No. 1, pp. 114–116, Jan 1970.

[12] S. P. Skobelev, "Radiation from an array of planar waveguides with slotted coupling elements,", *Soviet J. Commun. Technol. Electr.*, Vol. 42, No. 7, pp. 170–172, 1987.

[13] S. P. Skobelev, "Analysis and synthesis of an antenna array with sectoral partial radiation patterns," *Telecommun. Radio Eng.*, Vol. 45, pp. 116–119, Nov. 1990.

[14] E. C. Dufort, "Constrained feeds for limited scan arrays," *IEEE Trans.*, Vol. AP-26, pp. 407–413, May 1978.

[15] J. S. Herd, S. M. Duffy, D. D. Santiago and H. Steyskal, "Design considerations and results for an overlapped subarray radar antenna", *IEEE Aerospace Conference Digest*, Big Sky, Montana (2005).

[16] T. T. Hill, "Phased array feed systems, a survey," in *Phased Array Antennas: Proceeiings of the 1970 Phased Array Antenna Symposium*, Oliner and Knittel, Eds. Dedham, MA: Artech House Inc., 1972, pp. 197–209.

[17] R. J. Mailloux, *Phased Array Antenna Handbook*, 2nd edition. Dedham, MA: Artech House Publishing Co. 2000.

[18] G. V. Borgiotti, "An antenna for limited scan in one plane: Design criteria and numerical simulation," *IEEE Trans.*, Vol. AP-25, No.1, pp. 232–243, Jan. 1977.

[19] R. L. Fante, "Systems study of overlapped subarrayed scanning antennas," *IEEE Trans.*, Vol. AP-28, No. 5, pp. 668–679, Sept. 1980.

[20] H. L. Southall and D. T McGrath, "An experimental completely overlapped sybarray antenna," *IEEE Trans.*, AP-34, No. 4, pp. 465–474, April 1986.

[21] R. J. Mailloux, "A low-sidelobe partially overlapped constrained feed network for time-delayed subarrays," *IEEE Trans.*, Vol. AP-49, No. 2, pp. 280–291, Feb. 2001.

Author Biography

Robert J. Mailloux received the BS degree in electrical engineering from Northeastern University, Boston, MA in 1961 and the SM and Ph.D degrees at Harvard University, Cambridge MA in 1962 and 1965 respectively.

Currently he is a Research Professor at the University of Massachusetts, Amherst MA, and a member of the staff of Photonic Systems Inc.. He is a retired Senior Scientist at the Sensors Directorate, Air Force Research Laboratory, Hanscom Air Force Base, MA, and now working with University of Massachusetts on a research contract with the Air Force Research Laboratory. He has served as the Chief of the Antennas and Components Division, Rome Laboratory and as a physicist at the Air Force Cambridge Research Laboratory. He is the author or co-author of numerous journal articles, book chapters, 13 patents and the book *Phased Array Handbook* (Norwood MA; Artech House 1994) and co-editor of the book *History of Wireless*, with colleagues T. Sarkar, A. Oliner, M. Salazar-Palma and D. Gupta. His research interests are in the area of periodic structures and antenna arrays. He is a Life Fellow of the IEEE.

Dr. Mailloux was President of the Antennas and Propagation Society in 1983, and in 1992 he received the IEEE Harry Diamond Memorial Award. He received two IEEE AP-S Honorable Mention Best Paper Awards and recently received the IEEE Third Millennium Medal and received the AP-S Distinguished Technical Achievement Award in 2005. He is a member of Tau Beta Pi, Eta Kappa Nu, Sigma Xi, and Commission B of the International Scientific Radio Union. He has received four Air Force Research Laboratory "Best Paper" awards, the "Engineer of the Year" award, and is an AFRL Fellow. He has been a distinguished Lecturer for the Antenna and Propagation Society and is currently chairman of the AP-S/IEEE Press Liaison Committee.

Printed in the United States
by Baker & Taylor Publisher Services